Yuehui Chen and Ajith Abraham

Intelligent Systems Reference Library

Intelligent Systems Reference Library, Volume 2

Editors-in-Chief

Prof. Janusz Kacprzyk
Systems Research Institute
Polish Academy of Sciences
ul. Newelska 6
01-447 Warsaw
Poland
E-mail: kacprzyk@ibspan.waw.pl

Prof. Lakhmi C. Jain
University of South Australia
Adelaide
Mawson Lakes Campus
South Australia
Australia
E-mail: Lakhmi.jain@unisa.edu.au

Further volumes of this series can be found on our homepage: springer.com

Vol. 1. Christine L. Mumford and Lakhmi C. Jain (Eds.)
Computational Intelligence, 2009
ISBN 978-3-642-01798-8

Vol. 2. Yuehui Chen and Ajith Abraham
Tree-Structure Based Hybrid Computational Intelligence, 2009
ISBN 978-3-642-04738-1

Yuehui Chen and Ajith Abraham

Tree-Structure Based Hybrid Computational Intelligence

Theoretical Foundations and Applications

 Springer

Prof. Yuehui Chen
School of Information Science and Engineering
University of Jinan
Jiwei Road 106
Jinan 250022
P.R. China
E-mail: yhchen@ujn.edu.cn

Prof. Ajith Abraham
Machine Intelligence Research Labs (MIR Labs)
Scientific Network for Innovation and Research Excellence
P.O. Box 2259
Auburn, Washington 98071-2259
USA
E-mail: ajith.abraham@ieee.org

ISBN 978-3-642-26120-6   e-ISBN 978-3-642-04739-8

DOI 10.1007/978-3-642-04739-8

Intelligent Systems Reference Library   ISSN 1868-4394

Typeset & Cover Design: Scientific Publishing Services Pvt. Ltd., Chennai, India.

Printed in acid-free paper

9 8 7 6 5 4 3 2 1

springer.com

This book is dedicated to all our family members and colleagues around the world who supported us during the last several years.

This work is supported by Publishing Foundation Project of University of Jinan.

Preface

Computational intelligence is a well-established paradigm, where new theories with a sound biological understanding have been evolving. The current experimental systems have many of the characteristics of biological computers (brains in other words) and are beginning to be built to perform a variety of tasks that are difficult or impossible to do with conventional computers. In a nutshell, which becomes quite apparent in the light of the current research pursuits, the area is heterogeneous as being dwelled on such technologies as neurocomputing, fuzzy inference systems, artificial life, probabilistic reasoning, evolutionary computation, swarm intelligence and intelligent agents and so on.

Research in computational intelligence is directed toward building thinking machines and improving our understanding of intelligence. As evident, the ultimate achievement in this field would be to mimic or exceed human cognitive capabilities including reasoning, recognition, creativity, emotions, understanding, learning and so on. Even though we are a long way from achieving this, some success has been achieved in mimicking specific areas of human mental activity.

Recent research in computational intelligence together with other branches of engineering and computer science has resulted in the development of several useful intelligent paradigms. The integration of different learning and adaptation techniques, to overcome individual limitations and achieve synergetic effects through hybridization or fusion of some of these techniques, has in recent years contributed to a large number of new hybrid intelligent system designs.

Learning methods and approximation algorithms are fundamental tools that deal with computationally hard problems, in which the input is gradually disclosed over time. Both kinds of problems have a large number of applications arising from a variety of fields, such as function approximation and classification, algorithmic game theory, coloring and partitioning, geometric problems, mechanism design, network design, scheduling, packing and covering and real-world applications such as medicine, computational finance, and so on.

In this book, we illustrate Hybrid Computational Intelligence (HCI) framework and it applications for various problem solving tasks. Based on tree-structure based encoding and the specific function operators, the models can be flexibly constructed and evolved by using simple computational intelligence techniques. The main idea behind this model is the flexible neural tree, which is very adaptive, accurate and efficient. Based on the pre-defined instruction/operator sets, a flexible neural tree model can be created and evolved. The flexible neural tree could be evolved by using tree-structure based evolutionary algorithms with specific instructions. The fine tuning of the parameters encoded in the structure could be accomplished by using parameter optimization algorithms. The flexible neural tree method interleaves both optimizations. Starting with random structures and corresponding parameters, it first tries to improve the structure and then as soon as an improved structure is found, it fine tunes its parameters. It then goes back to improving the structure again and, provided it finds a better structure, it again fine tunes the rules' parameters. This loop continues until a satisfactory solution is found or a time limit is reached.

This volume is organized into 6 Chapters and the main contributions are detailed below:

Chapter 1 provides a gentle introduction to some of the key paradigms in computational intelligence namely evolutionary algorithms and its variants, swarm intelligence, artificial neural networks, fuzzy expert systems, probabilistic computing and hybrid intelligent systems.

Chapter 2 exhibits the flexible neural tree algorithm development and is first illustrated in some function approximation problems and also in some real world problems like intrusion detection, exchange rate forecasting, face recognition, cancer detection and protein fold recognition. Further the multi-input multi-output flexible neural tree algorithm is introduced and is illustrated for some problem solving. Finally an ensemble of flexible neural trees is demonstrated for stock market prediction problem.

Chapter 3 depicts three different types of hierarchical architectures. First the design and implementation of hierarchical radial basis function networks are illustrated for breast cancer detection and face recognition. Further, the development of hierarchical B-spline networks is demonstrated for breast cancer detection and time series prediction. Finally, hierarchical wavelet neural networks are presented for several function approximation problems.

Building a hierarchical fuzzy system is a difficult task. This is because the user has to define the architecture of the system (the modules, the input variables of each module, and the interactions between modules), as well as the rules of each modules. **Chapter 4** demonstrates a new encoding and an automatic design method for the hierarchical Takagi-Sugeno fuzzy inference system with some simulation results related to system identification and time-series prediction problems.

Can we evolve a symbolic expression that can be represented as a meaningful expression, i.e., a differential equation or a transfer function and it can be easily addressed by using traditional techniques? **Chapter 5** exhibits a new representation scheme of the additive models, by which the linear and nonlinear system identification problems are addressed by using automatic evolutionary design procedure. First a gentle introduction to tree structural representation and calculation of the additive tree models is provided. Further an hybrid algorithm for evolving the additive tree models and some simulation results for the prediction of chaotic time series, the reconstruction of polynomials and the identification of the linear/nonlinear system is demonstrated.

Chapter 6 summarizes the concept of hierarchical hybrid computational intelligence framework introduced in this book and also provides some future research directions.

We are very much grateful to Dr. Thomas Ditzinger (Springer Engineering Inhouse Editor, Professor Janusz Kacprzyk (Editor- in-Chief, Springer *Intelligent Systems Reference Library* Series) and Ms. Heather King (Editorial Assistant, Springer Verlag, Heidelberg) for the editorial assistance and excellent cooperative collaboration to produce this important scientific work. We hope that the reader will share our joy and will find it useful!

Yuehui Chen and Ajith Abraham*
School of Information Science and Engineering,
University of Jinan, Jiwei Road 106, Jinan 250022,
Peoples Republic of China
http://cilab.ujn.edu.cn
Email: yhchen@ujn.edu.cn

*Machine Intelligence Research Labs (MIR Labs)
Scientific Network for Innovation and Research Excellence
P.O. Box 2259, Auburn, Washington 98071, USA
http://www.mirlabs.org
http://www.softcomputing.net
email: ajith.abraham@ieee.org

Contents

Part II: Flexible Neural Trees

Part III: Hierarchical Neural Networks

Part IV: Hierarchical Fuzzy Systems

Part V: Reverse Engineering of Dynamical Systems

Part VI: Conclusions and Future Research

Part I

Foundations of Computational Intelligence

1

Foundations of Computational Intelligence

1.1 Introduction

The field of computational intelligence has evolved with the objective of developing machines that can think like humans. Computational intelligence is a well-established paradigm, where new theories with a sound biological understanding have been evolving. The current experimental systems have many of the characteristics of biological computers (brains in other words) and are beginning to be built to perform a variety of tasks that are difficult or impossible to do with conventional computers. To name a few, we have microwave ovens, washing machines and digital camera that can figure out on their own what settings to use to perform their tasks optimally with reasoning capability, make intelligent decisions and learn from experience. As usual, defining computational intelligence is not an easy task. In a nutshell, which becomes quite apparent in light of the current research pursuits, the area is heterogeneous as being dwelled on such technologies as neural networks, fuzzy systems, evolutionary computation, artificial life, multi-agent systems and probabilistic reasoning. The recent trend is to integrate different components to take advantage of complementary features and to develop a synergistic system. Hybrid architectures like neuro-fuzzy systems, evolutionary-fuzzy systems, evolutionary-neural networks, evolutionary neuro-fuzzy systems etc. are widely applied for real world problem solving.

This Chapter provides a gentle introduction to some of the key paradigms in computational intelligence namely evolutionary algorithms and its variants, swarm intelligence, artificial neural networks, fuzzy expert systems, probabilistic computing and hybrid intelligent systems.

1.2 Evolutionary Algorithms

Evolution can be viewed as a search process capable of locating solutions to problems offered by an environment. Therefore, it is quite natural to look for

Y. Chen, A. Abraham.: Tree-Struc. Based Hybrid Com. Intelligence, ISRL 2, pp. 3–36.
springerlink.com

an algorithmic description of evolution that can be used for problem solving. Such an algorithmic view has been discussed even in philosophy. Those iterative (search and optimization) algorithms developed with the inspiration of the biological process of evolution are termed evolutionary algorithms (EAs). They are aimed basically at problem solving and can be applied to a wide range of domains, from planning to control. Evolutionary computation (EC) is the name used to describe the field of research that embraces all evolutionary algorithms. The basic idea of the field of evolutionary computation, which came onto the scene about the 1950s/1960s, has been to make use of the powerful process of natural evolution as a problem-solving paradigm, usually by simulating it on a computer. The original three mainstreams of EC are genetic algorithms (GAs), evolution strategies (ES), and evolutionary programming (EP) [1][2]. Despite some differences among these approaches, all of them present the basic features of an evolutionary process as proposed by the Darwinian theory of evolution.

A standard evolutionary algorithm is illustrated as follows:

- A population of individuals that reproduce with inheritance. Each individual represents or encodes a point in a search space of potential solutions to a problem. These individuals are allowed to reproduce (sexually or asexually), generating offspring that carry some resemblance with their parents;
- Genetic variation. Offspring are prone to genetic variation through mutation, which alters their genetic makeup;
- Natural selection. The evaluation of individuals in their environment results in a measure of adaptability, quality, or fitness value to be assigned to them. A comparison of individual fitnesses will lead to a competition for survival and reproduction in the environment, and there will be a selective advantage for those individuals of higher fitness [306].

The standard evolutionary algorithm is a generic, iterative and probabilistic algorithm that maintains a population P of N individuals, $P = x_1, x_2, , x_N$, at each iteration t (for simplicity of notation the iteration index t was suppressed). Each individual corresponds to (represents or encodes) a potential solution to a problem that has to be solved. An individual is represented using a data structure. The individuals x_i, $i = 1, , N$, are evaluated to give their measures of adaptability to the environment, or fitness. Then, a new population, at iteration $t + 1$, is generated by selecting some (usually the most fit) individuals from the current population and reproducing them, sexually or asexually. If employing sexual reproduction, a genetic recombination (crossover) operator may be used. Genetic variations through mutation may also affect some individuals of the population, and the process iterates. The completion of all these steps: reproduction, genetic variation, and selection, constitutes what is called a generation. An initialization procedure is used to generate the initial population of individuals. Two parameters p_c and p_m correspond to the genetic recombination and variation probabilities, and will be further discussed.

Note that all evolutionary algorithms involve the basic concepts common to every algorithmic approach to problem solving:

- representation (data structures);
- definition of an objective; and
- specification of an evaluation function (fitness function).

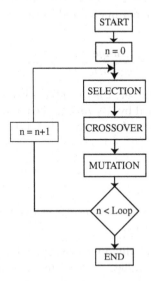

Fig. 1.1 A flowchart of simple genetic algorithm

Genetic algorithms (GAs) are globally stochastic search technique that emulates the laws of evolution and genetics to try to find optimal solutions to complex optimization problems. GAs are theoretically and empirically proven to provide to robust search in complex spaces, and they are widely applied in engineering, business and scientific circles. The general flowchart of GA is presented in Figure 1.1.

GAs are different from more normal optimization and search procedures in different ways:

- GAs work with a coding of the parameter set, not the parameter themselves.
- GAs search from a population of points, not a single point.
- GAs use objective function information, not derivatives or other auxiliary knowledge, but with modifications they can exploit analytical gradient information if it is available.
- GAs use probabilistic transition rules, not deterministic rules.

Coding and Decoding

Coding refers to the representation of the parameter used in the optimization problem. The usually used coding methods in GAs are base-2, base-10 and floating-point coding methods. In a base-2 representation, alleles (values in the position, genes on the chromosome) are 0 and 1. In base-10, the alleles take on integer values between 0 and 9. In floating-point representation, the alleles are real-valued number. In base-2 and base-10 representations, the relationship between the real value of a parameter and its integer representation can be expressed by:

$$x = a + \bar{x}\frac{range}{resolution} \tag{1.1}$$

where x is the real value of the parameter, $\bar{x}$ is the integer value corresponding to the x, a is the lowest value assumed by $\bar{x}$, $range$ is the interval of definition of the parameters, and $resolution$ is the number that take in account the number of bits used, i.e., $2^{number\ of\ bits} - 1$.

Genetic Operators

A simple genetic algorithm that yields good results in many practical problems consists of three genetic operators:

- $Reproduction$ is a process in which individual strings are copied according to their objective or fitness function values. Fitness function can be imagined as some measure of profit, utility, or goodness to be optimized. For example, in curve fitting problem, the fitness function can be mean square error:

$$Fitness = \frac{1}{n}\sum_{i=1}^{n}(y_i - f(y_i, a_i))^2 \tag{1.2}$$

  where y_i is the experimental data, $f(y_i, a_i)$ is the function chosen as model and a_i are the model parameters to be optimized by GA. When GA is used to optimize an adaptive controller, the error and change in error information can be taken account for the designing of a proper fitness function. In general, reproduction operator guarantee survival of the better individual to the next generation with a higher probability, which is an artificial version of natural selection.
- $Crossover$ is a partial exchange of the genetic content between couples of members of the population. This task can be done in several different ways and it also depends on the representation scheme chosen. In integer representation, the simple way to do it, is to choose a random value with a uniform distribution as $[1, length\ of\ chromosome]$. This number represents a marker inside the two strings of bits representing the couple of chromosomes. It cuts both the chromosomes into two parts. Then, the left

or the right parts of the two chromosomes are swapped. This occurs in such a way that both the two new chromosomes will contain a part of the genetic information of both the parents. In the floating-point representation, the crossover should be realized by:

$$new_1 = a \cdot old_1 + (1 - a) \cdot old_2, \qquad (1.3)$$
$$new_2 = (1 - a) \cdot old_1 + a \cdot old_2, \qquad (1.4)$$

where new_1 and new_2 are the chromosomes after the crossover, old_1 and old_2 are the chromosomes before the crossover, a is a random number with uniform distribution in $[0,1]$.

- *Mutation* is needed because, even through reproduction and crossover effectively search and recombine extant notions, occasionally they may become overzealous and lose some potentially useful genetic materials. In GA, the mutation operator protects against such an irrecoverable loss. In other words, mutation tries to escape from a local maximum or minimum of the fitness function, and it seeks to explore other areas of the search space in order to find a global maximum or minimum of the fitness function. In integer representation, the mutation of gene in a position of the chromosome is randomly changed form one integer to another. In floating-point representation, mutation will randomly change the value of the chromosome within a range of definition.

1.2.1 Genetic Programming

Genetic Programming (GP) technique provides a framework for automatically creating a working computer program from a high-level problem statement of the problem [30]. Genetic programming achieves this goal of automatic programming by genetically breeding a population of computer programs using the principles of Darwinian natural selection and biologically inspired operations. The main difference between genetic programming and genetic algorithms is the representation of the solution. Genetic programming creates computer programs in the LISP or scheme computer languages as the solution. LISP is an acronym for LISt Processor and was developed by John McCarthy in the late 1950s. Unlike most languages, LISP is usually used as an interpreted language. This means that, unlike compiled languages, an interpreter can process and respond directly to programs written in LISP. The main reason for choosing LISP to implement GP is due to the advantage of having the programs and data have the same structure, which could provide easy means for manipulation and evaluation.

GP is the extension of evolutionary learning into the space of computer programs. In GP the individual population members are not fixed length character strings that encode possible solutions to the problem at hand, they are programs that, when executed, are the candidate solutions to the problem. These programs are expressed in genetic programming as parse trees,

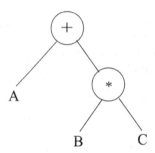

Fig. 1.2 A simple tree structure of GP

rather than as lines of code. Example, the simple program $a + b * c$ would be represented as shown in Figure 1.2. The terminal and function sets are also important components of genetic programming. The terminal and function sets are the alphabet of the programs to be made. The terminal set consists of the variables and constants of the programs (example, A,B and C in Figure 1.2).

The most common way of writing down a function with two arguments is the infix notation. That is, the two arguments are connected with the operation symbol between them as follows:

$$A + B$$

A different method is the prefix notation. Here the operation symbol is written down first, followed by its required arguments.

$$+AB$$

While this may be a bit more difficult or just unusual for human eyes, it opens some advantages for computational uses. The computer language LISP uses symbolic expressions (or S-expressions) composed in prefix notation. Then a simple S-expression could be

$$(operator, argument)$$

where operator is the name of a function and argument can be either a constant or a variable or either another symbolic expression as shown below:

$$(operator, argument(operator, argument)(operator, argument))$$

A parse tree (Figure 1.3) is a structure that develops the interpretation of a computer program. Functions are written down as nodes, their arguments as leaves. A subtree is the part of a tree that is under an inner node of this tree. If this tree is cut out from its parent, the inner node becomes a root node and the subtree is a valid tree of its own.

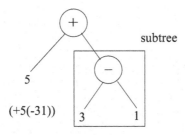

Fig. 1.3 Illustration of a parse tree and a subtree

There is a close relationship between these parse trees and S-expression; in fact these trees are just another way of writing down expressions. While functions will be the nodes of the trees (or the operators in the S-expressions) and can have other functions as their arguments, the leaves will be formed by terminals, that is symbols that may not be further expanded. Terminals can be variables, constants or specific actions that are to be performed. The process of selecting the functions and terminals that are needed or useful for finding a solution to a given problem is one of the key steps in GP.

Evaluation of these structures is straightforward. Beginning at the root node, the values of all sub-expressions (or subtrees) are computed, descending the tree down to the leaves. GP procedure could be summarized as follows:

- Generate an initial population of random compositions of the functions and terminals of the problem;
- Compute the fitness values of each individual in the population;
- Using some selection strategy and suitable reproduction operators produce offsprings;
- Procedure is iterated until the required solution is found or the termination conditions have reached (specified number of generations).

The creation of an offspring from the crossover operation is accomplished by deleting the crossover fragment of the first parent and then inserting the crossover fragment of the second parent. The second offspring is produced in a symmetric manner. A simple crossover operation is illustrated in Figure 1.4. In GP the crossover operation is implemented by taking randomly selected sub trees in the individuals and exchanging them.

Mutation is another important feature of genetic programming. Two types of mutations are commonly used. The simplest type is to replace a function or a terminal by a function or a terminal respectively. In the second kind an entire subtree can replace another subtree. Figure 1.5 explains the concept of mutation:

GP requires data structures that are easy to handle and evaluate and robust to structural manipulations. These are among the reasons why the class of S-expressions was chosen to implement GP. The set of functions and terminals that will be used in a specific problem has to be chosen carefully. If the set of

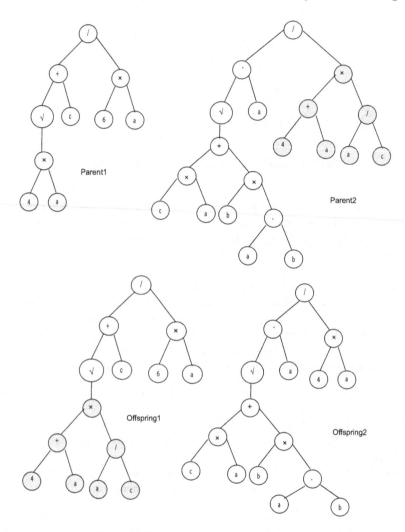

Fig. 1.4 Illustration of crossover operator

functions is not powerful enough, a solution may be very complex or not to be
found at all. Like in any evolutionary computation technique, the generation
of first population of individuals is important for successful implementation
of GP. Some of the other factors that influence the performance of the algorithm are the size of the population, percentage of individuals that participate
in the crossover/mutation, maximum depth for the initial individuals and the
maximum allowed depth for the generated offspring etc. Some specific advantages of genetic programming are that no analytical knowledge is needed
and still could get accurate results. GP approach does scale with the problem

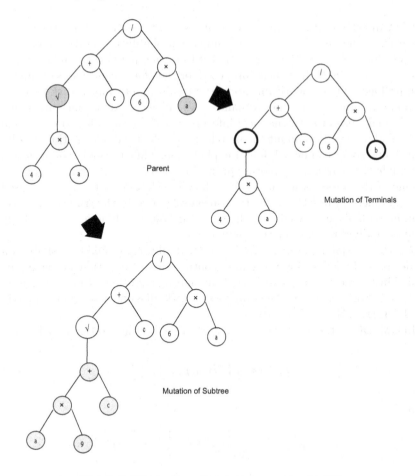

Fig. 1.5 Illustration of mutation operator in GP

size. GP does impose restrictions on how the structure of solutions should be formulated.

1.2.2 Estimation of Distribution Algorithm

Estimation of distribution algorithms (EDAs) are a novel class of evolutionary optimization algorithms that were developed as a natural alternative to genetic algorithms [15][16][17][18][19][20]. The principal advantages of EDAs over genetic algorithms are the absence of multiple parameters to be tuned (e.g. crossover and mutation probabilities) and the expressiveness and transparency of the probabilistic model that guides the search process. In addition, EDAs have been proven to be better suited to some applications than GAs, while achieving competitive and robust results in the majority of tackled problems.

EDA directly extracts the global statistical information about the search space from the search so far and builds a probabilistic model of promising solutions. New solutions are sampled from the model thus built. Several EDAs [17][18][19][20] have been proposed for the global continuous optimization problem. These algorithms are very promising, but much work needs to be done to improve their performances. An efficient evolutionary algorithm should make use of both the local information of solutions found so far and the global information about the search space. The local information of solutions found so far can be helpful for exploitation, while the global information can guide the search for exploring promising areas.

One of the major issues in EDAs is how to select parents. A widely-used selection method in EDA is the truncation selection. In the truncation selection, individuals are sorted according to their objective function values. Only the best individuals are selected as parents.

Another major issue in EDAs is how to build a probability distribution model p(x). In EDAs for the global continuous optimization problem, the probabilistic model p(x) can be a Gaussian distribution, a Gaussian mixture, a histogram, or a Gaussian model with diagonal covariance matrix (GM/DCM) [15][16][17][18][19][20].

In GM/DCM, the joint density function of the k-th generation is as follows:

$$p_k(x) = \prod_{i=1}^{n} N(x_i; \mu_i^k, \sigma_i^k) \qquad (1.5)$$

where

$$N(x_i; \mu_i^k, \sigma_i^k) = \frac{1}{\sqrt{2\pi}\sigma_i} exp(-\frac{1}{2}(\frac{x_i - \mu_i}{\sigma_i})^2) \qquad (1.6)$$

In the above equation, the n-dimensional joint probability distribution is factorized as a product of n univariate and independent normal distributions. There are two parameters for each variable required to be estimated in the k-th generation: the mean, μ_i^k, and the standard deviation, σ_i^k. They can be estimated as follows:

$$\hat{\mu}_i^k = \bar{x}_i^k = \frac{1}{M} \sum_{j=1}^{M} x_{ji}^k \qquad (1.7)$$

$$\hat{\sigma}_i^k = \sqrt{\frac{1}{M} \sum_{j=1}^{M} (x_{ji}^k - \bar{x}_i^k)^2} \qquad (1.8)$$

1.2.3 Population-Based Incremental Learning

Population-based incremental learning (PBIL) combines elements from evolutionary computation (EC) and reinforcement learning (RL) [24][25]. PBIL is a population-based stochastic search where the population is essentially a random sample based on an estimated probability distribution for each variable. So, in reality the population does not exist as it does in traditional EC. After a sample is generated, the best is retained and the probability model for each variable is updated to reflect the belief regarding the structure of the best solution. This is accomplished according to a similar update rule as used in RL. The result is a statistical approach to evolutionary computation.

An evolutionary algorithm's population can be thought of as representing an estimated probability distribution over the possible values for each gene. In PBIL the population is replaced by a $d \times c$ dimensional probability matrix $M := (m_{ij})_{d \times c}$ which corresponds to a probability distribution over possible values for each element (d is the problem dimensionality each having c variables). For example, if a binary problem is under consideration then a solution $B := (b_{ij})_{d \times c}$ where $b_{i,j} \in 0, 1$ and so each $m_{ij} \in [0, 1]$ corresponds to the probability of $b_{ij} = 1$.

Learning in PBIL consists of utilizing the current M to generate a set G_1 of k samples. These samples are evaluated according to the fitness function for the given problem and the best sample $B^* = (b_{ij})_{d \times c} \in 0, 1$ is maintained. Then, the probability distributions represented in M are updated by increasing the probability of generating solutions similar to B^*. The update rule to accomplish this is similar to that found in learning vector quantization,

$$M_t = (1 - \alpha)M_{t-1} + \alpha B^* \tag{1.9}$$

where $0 < \alpha < 1$ represents a user-defined learning rate and the subscript $t \geq 1$ corresponds to the current iteration of PBIL. Without prior information $(m_{ij}) = 0.5$.

Another contrast to evolutionary computation is the lack of a crossover operator or selection mechanism, instead the values in M are mutated once per iteration. During this phase a small random value is added or subtracted from a random subset of the values in M. Furthermore, since at each iteration a new subset of samples is generated and only the best is maintained then no selection mechanism is required.

The pseudocode for PBIL is presented in Algorithm 1. It assumes constants to control the maximum number of iterations and samples, x and k, respectively.

Initially, in line 2 we let $M = (m_{ij}) = 0.5$ to reflect the lack of a priori information regarding the probability distribution of each variable. In line 5 we generate the k samples using the current probability matrix and select the best sample (w.r.t. some user-defined criteria) in line 7. Matrix M is updated in line 9 using the best sample to guide the direction of probability update. Finally, lines 11-15 probabilistically perform the mutation rule. It

---

Algorithm 1 Population-Based Incremental Learning [24][25]

---

01. Initialize probabilities
02. $M_0 = (m_{ij}) = 0.5$
03. for $t = 1$ to ω
04. Generate samples
05. $G_1 = generate\_samples(k, M_{t-1})$
06. Find best sample
07. $B^* = slect_best(B^* \cup G_1)$
08. Update M
09. $M_t = (1 - \alpha)M_{t-1} + \alpha B^*$
10. Mutate probability vector
10. for $i = 0, \ldots, d$ and $j = 0, \ldots, c$ do
12.   if random(0,1)$<\beta$ then
13. $m_{ij} = (1 - \gamma)m_{ij} + \gamma \cdot random(0 or 1)$
14.   end if
15. end for
16. end for

---

has been shown that for a given discrete search space PBIL will converge to a local optima [26][27]. PBIL algorithms for continuous spaces have also been explored (for examples see [28][29]).

1.2.4   Probabilistic Incremental Program Evolution

Probabilistic Incremental Program Evolution (PIPE) combines probability vector coding of program instructions, population based incremental learning [222][223], and tree-coded programs [219][220][221][226]. PIPE iteratively generates successive populations of functional programs according to an adaptive probability distribution, represented as a Probabilistic Prototype Tree (PPT), over all possible programs. Each iteration uses the best program to refine the distribution. Thus, the structures of promising individuals are learned and encoded in PPT.

Instructions and Programs

In PIPE, programs are made of instructions from an instruction set $S = \{I_1, I_2, \ldots, I_n\}$ with n instructions. Instructions are user-defined and problem dependent. Each instruction is either a function or a terminal. Instruction set S therefore consists of a function set $F = \{f_1, f_2, \ldots, f_k\}$ with k functions and a terminal set $T = \{t_1, t_2, \ldots, t_l\}$ with l terminals, where $n = k + l$ holds.

Programs are encoded in n-ary trees, with n being the maximal number of function arguments. Each non-leaf node encodes a function from F and each leaf node a terminal from T. The number of subtrees each node has corresponds to the number of arguments of its function. Each argument is calculated by a subtree. The trees are parsed depth first from left to right.

Probabilistic Prototype Tree (PPT)

The PPT stores the knowledge gained from experiences with programs and guides the evolutionary search. It holds random constants and the probability distribution over all possible programs that can be constructed from a predefined instruction set. The PPT is generally a complete n-ary tree with infinitely many nodes, where n is the maximal number of function arguments.

Each node N_j in PPT, with $j > 0$ contains a random constant R_j and a variable probability vector $\overrightarrow{P_j}$. Each $\overrightarrow{P_j}$ has n components, where n is the number of instructions in instruction set S. Each component $P_j(I)$ of $\overrightarrow{P_j}$ denotes the probability of choosing instruction $I \in S$ at node N_j. Each vector $\overrightarrow{P_j}$ is initialized as follows:

$$P_j(I) = \frac{P_T}{l}, \forall I : I \in T \tag{1.10}$$

$$P_j(I) = \frac{1 - P_T}{k}, \forall I : I \in F, \tag{1.11}$$

where l is the total number of terminals in T, k is the total number of functions in F, and P_T is initially user-defined constant probability for selecting an instruction from T.

1.2.4.1 Program Generation, Growing and Pruning

Programs are generated according to the probability distribution stored in the PPT. To generate a program P_{ROG} from PPT, an instruction $I \in S$ is selected with probability $P_j(I)$ for each accessed node N_j of PPT. Nodes are accessed in a depth-first way, starting at the root node and traversing PPT from left to right.

A complete PPT is infinite, and each PPT node holds a probability for each instruction, a random constant, and n pointers to following nodes, where n is PPT's arity. Therefore, A large PPT is memory intensive. To reduce memory requirements, it is thus possible to incrementally grow and prune the PPT.

On one hand, it is useful to grow the PPT on demand in order to create a variety of programs. Initially the PPT contains only the root node. Additional nodes are created with each program that accesses non-existing nodes during its generation. On the other hand, apart from reducing memory requirements, pruning also helps to discard the elements of probability distribution that have become irrelevant over time. PPT subtrees attached to nodes that contain at least one probability vector component above a threshold T_P can be pruned. If T_P is set to a sufficiently high value (e.g., $T_P = 0.99999$) only parts of the PPT will be pruned that have a very low probability of being accessed. In case of functions, only those subtrees should be pruned that are not required as function arguments. Figure 1.6 illustrates the relation between the prototype tree and a possible program tree.

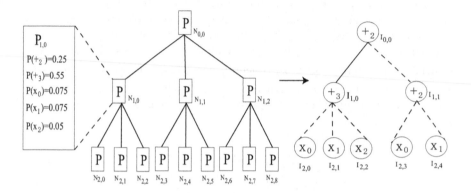

Fig. 1.6 Example of node $N_{1,0}$'s instruction probability vector $P_{1,0}$ (left). Probabilistic prototype tree PPT (middle). Possible extracted program P_{ROG}, at the time of creation of instruction $I_{1,0}$, the dashed part of P_{ROG} did not exist yet (right).

Fitness Functions

Similar to the other evolutionary algorithms, PIPE uses a problem-dependent and user-defied fitness function. A fitness function maps programs to scalar, real-valued fitness values that reflect the programs' performances on a given task. Firstly PIPE's fitness functions should be seen as error measures, i.e., $MeanSquaredError(MSE)E$ or Root Mean Squared Error (RMSE). A secondary non-user-defined objective for which PIPE always optimizes programs is the program size as measured by number of nodes. Among programs with equal fitness smaller ones are always preferred. This objective constitutes PIPE's built-in Occam's razor.

Learning Algorithm

PIPE combines two forms of learning: Generation-Based Learning (GBL) and Elitist Learning (EL). GBL is PIPE's main learning algorithm. EL's purpose is to make the best program found so far as an attractor. PIPE executes:

GBL
REPEAT

> with probability P_{el} DO EL
> otherwise DO GBL

UNTIL termination criterion is reached

Here P_{el} is a user-defined constant in the interval $[0,1]$.

Generation-Based Learning

Step 1. **Creation of Program Population.** A population of programs P_{ROGj} ($0 < j \leq PS$; PS is population size) is generated using the prototype tree PPT, as described above. The PPT is grown on demand.

Step 2. **Population Evaluation.** Each program P_{ROG_j} of the current population is evaluated on the given task and assigned a fitness value $FIT(P_{ROG_j})$ according to the predefined fitness function. The best program of the current population (the one with the smallest fitness value) is denoted P_{ROG_b}. The best program found so far (elitist) is preserved in $P_{ROG}{}^{el}$.

Step 3. **Learning from Population.** Prototype tree probabilities are modified such that the probability $P(P_{ROG_b})$ of creating P_{ROG_b} increases. This procedure is called adapt PPT towards(Progb). This is implemented as follows. First $P(P_{ROG_b})$ is computed by looking at all PPT nodes N_j used to generate P_{ROG_b}:

$$P(P_{ROG_b}) = \prod_{j:N_j \ used \ to \ generate \ P_{ROG_b}} P_j\left(I_j(P_{ROG_b})\right) \qquad (1.12)$$

where $I_j(P_{ROG_b})$ denotes the instruction of program P_{ROG_b} at node position j. Then a target probability P_{TARGET} for P_{ROG_b} is calculated:

$$P_{TARGET} = P(P_{ROG_b}) + (1 - P(P_{ROG_b})) \cdot lr \cdot \frac{\varepsilon + FIT(P_{ROG}{}^{el})}{\varepsilon + FIT(P_{ROG_b})} \quad (1.13)$$

Here lr is a constant learning rate and ε a positive user-defined constant. Given P_{TARGET}, all single node probabilities $P_j(I_j(P_{ROG_b}))$ are increased iteratively:

REPEAT:

$$P_j\left(I_j(P_{ROG_b})\right) = P_j\left(I_j(P_{ROG_b})\right) + c^{lr} \cdot lr \cdot (1 - P_j\left(I_j(P_{ROG_b})\right)) \quad (1.14)$$

UNTIL $P(P_{ROG_b}) \geq P_{TARGET}$

where c^{lr} is a constant influencing the number of iterations. The smaller c^{lr} the higher the approximation precision of P_{TARGET} and the number of required iterations. Setting $c^{lr} = 0.1$ turned out to be a good compromise between precision and speed. And then all adapted vectors $\vec{P_j}$ are renormalized.

Step 4. **Mutation of Prototype Tree.** All probabilities $P_j(I)$ stored in nodes N_j that were accessed to generate program P_{ROG_b} are mutated with a probability P_{M_p}:

$$P_{M_p} = \frac{P_M}{n \cdot \sqrt{|P_{ROG_b}|}} \tag{1.15}$$

where the user-defined parameter P_M defines the overall mutation probability, n is the number of instructions in instruction set S and $|P_{ROG_b}|$ denotes the number of nodes in program P_{ROG_b}. Selected probability vector components are then mutated as follows:

$$P_j(I) = P_j(I) + mr \cdot (1 - P_j(I)) \tag{1.16}$$

where mr is the mutation rate, another user-defined parameter. Also all mutated vectors $\vec{P_j}$ are renormalized.

Step 5. **Prototype Tree Pruning.** At the end of each generation the prototype tree is pruned, as described in Section III.B.

Step 6. **Termination Criteria.** Repeat above procedure until a fixed number of program evaluations is reached or a satisfactory solution is found.

Elitist Learning
Elitist learning focuses search on previously discovered promising parts of the search space. The PPT is adapted towards the elitist program $P_{ROG}{}^{el}$. This is realized by replacing the P_{ROG_b} with $P_{ROG}{}^{el}$ in learning from population in Step 3. It is particularly useful with small population sizes and works efficiently in the case of noise-free problems.

1.3 Swarm Intelligence

Swarm Intelligence (SI) is an innovative distributed intelligent paradigm for solving optimization problems that originally took its inspiration from the biological examples by swarming, flocking and herding phenomena in vertebrates. Swarm intelligence is aimed at collective behaviour of intelligent agents in decentralized systems. Although there is typically no centralized control dictating the behaviour of the agents, local interactions among the agents often cause a global pattern to emerge. Ant Colony Optimization (ACO), have already been applied successfully to solve several engineering optimization problems. Swarm models are population-based and the population is initialised with a population of potential solutions. These individuals are then manipulated (optimised) over many several iterations using several heuristics inspired from the social behaviour of insects in an effort to find the optimal solution. Ant colony algorithms are inspired by the behavior of natural ant colonies, in the sense that they solve their problems by multi agent cooperation using indirect communication through modifications in the environment. Ants release a certain amount of pheromone (hormone) while walking, and each ant prefers (probabilistically) to follow a direction, which is rich of pheromone. This simple behavior explains why ants are able to adjust to changes in the environment, such as optimizing shortest path to a food source or a nest. In ACO, ants use information collected during past simulations to direct their search and this information is available and modified through the environment. Recently ACO algorithms have also been used for clustering data sets.

1.3.1   Particle Swarm Optimization

Particle Swarm Optimization (PSO) incorporates swarming behaviors observed in flocks of birds, schools of fish, or swarms of bees, and even human social behavior, from which the idea is emerged [3, 7, 13]. PSO is a population-based optimization tool, which could be implemented and applied easily to solve various function optimization problems, or the problems that can be transformed to function optimization problems. As an algorithm, the main strength of PSO is its fast convergence, which compares favorably with many global optimization algorithms like Genetic Algorithms (GA) [6], Simulated Annealing (SA) [12, 14] and other global optimization algorithms. For applying PSO successfully, one of the key issues is finding how to map the problem solution into the PSO particle, which directly affects its feasibility and performance [312].

Canonical Particle Swarm Optimization

The canonical PSO model consists of a swarm of particles, which are initialized with a population of random candidate solutions. They move iteratively through the d-dimension problem space to search the new solutions, where the fitness, f, can be calculated as the certain qualities measure. Each particle has a position represented by a position-vector $\mathbf{x}_i$ (i is the index of the particle), and a velocity represented by a velocity-vector $\mathbf{v}_i$. Each particle remembers its own best position so far in a vector $\mathbf{x}_i^{\#}$, and its j-th dimensional value is $x_{ij}^{\#}$. The best position-vector among the swarm so far is then stored in a vector $\mathbf{x}^*$, and its j-th dimensional value is x_j^*. During the iteration time t, the update of the velocity from the previous velocity to the new velocity is determined by Eq.(1.17). The new position is then determined by the sum of the previous position and the new velocity by Eq.(1.18).

$$v_{ij}(t+1) = wv_{ij}(t) + c_1 r_1(x_{ij}^{\#}(t) - x_{ij}(t)) + c_2 r_2(x_j^*(t) - x_{ij}(t)). \quad (1.17)$$

$$x_{ij}(t+1) = x_{ij}(t) + v_{ij}(t+1). \quad (1.18)$$

where w is called as the inertia factor, r_1 and r_2 are the random numbers, which are used to maintain the diversity of the population, and are uniformly distributed in the interval $[0,1]$ for the j-th dimension of the i-th particle. c_1 is a positive constant, called as coefficient of the self-recognition component, c_2 is a positive constant, called as coefficient of the social component. From Eq.(1.17), a particle decides where to move next, considering its own experience, which is the memory of its best past position, and the experience of its most successful particle in the swarm. In the particle swarm model, the particle searches the solutions in the problem space with a range $[-s, s]$ (If the range is not symmetrical, it can be translated to the corresponding symmetrical range.) In order to guide the particles effectively in the search space, the maximum moving distance during one iteration must be clamped in between the maximum velocity $[-v_{max}, v_{max}]$ given in Eq.(1.19):

$$v_{ij} = sign(v_{ij})min(|v_{ij}|, v_{max}). \qquad (1.19)$$

The value of v_{max} is $p \times s$, with $0.1 \leq p \leq 1.0$ and is usually chosen to be s, i.e. $p = 1$. The pseudo-code for particle swarm optimization algorithm is illustrated in Algorithm 2.

Algorithm 2 Particle Swarm Optimization Algorithm

---

01. Initialize the size of the particle swarm n, and other parameters.
02. Initialize the positions and the velocities for all the particles randomly.
03. While (the end criterion is not met) do
04. $t = t + 1$;
05.   Calculate the fitness value of each particle;
06. $\mathbf{x}^* = argmin_{i=1}^{n}(f(\mathbf{x}^*(t-1)), f(\mathbf{x}_1(t)), f(\mathbf{x}_2(t)), \cdots, f(\mathbf{x}_i(t)), \cdots, f(\mathbf{x}_n(t)))$;
07.   For $i=$ 1 to n
08. $\mathbf{x}_i^{\#}(t) = argmin_{i=1}^{n}(f(\mathbf{x}_i^{\#}(t-1)), f(\mathbf{x}_i(t))$;
09.   For $j = 1$ to $Dimension$
10.   Update the j-th dimension value of $\mathbf{x}_i$ and $\mathbf{v}_i$
10.   according to Eqs.(1.17), (1.18), (1.19);
12.   Next j
13.   Next i
14. End While.

The end criteria are usually one of the following:

- Maximum number of iterations: the optimization process is terminated after a fixed number of iterations, for example, 1000 iterations.
- Number of iterations without improvement: the optimization process is terminated after some fixed number of iterations without any improvement.
- Minimum objective function error: the error between the obtained objective function value and the best fitness value is less than a pre-fixed anticipated threshold.

The Parameters of PSO

The role of inertia weight w, in Eq.(1.17), is considered critical for the convergence behavior of PSO. The inertia weight is employed to control the impact of the previous history of velocities on the current one. Accordingly, the parameter w regulates the trade-off between the global (wide-ranging) and local (nearby) exploration abilities of the swarm. A large inertia weight facilitates global exploration (searching new areas), while a small one tends to facilitate local exploration, i.e. fine-tuning the current search area. A suitable value for the inertia weight w usually provides balance between global and local exploration abilities and consequently results in a reduction of the number of iterations required to locate the optimum solution. Initially, the inertia weight is set as a constant. However, some experiment results indicates that it is better to initially set the inertia to a large value, in order to promote global exploration of the search space, and gradually decrease it to get more refined solutions [11]. Thus, an initial value around 1.2 and gradually reducing towards 0 can be considered as a good choice for w. A better method is to use some adaptive approaches (example: fuzzy controller), in which the parameters can be adaptively fine tuned according to the problems under consideration [9], [10].

The parameters c_1 and c_2, in Eq.(1.17), are not critical for the convergence of PSO. However, proper fine-tuning may result in faster convergence and alleviation of local minima. As default values, usually, $c_1 = c_2 = 2$ are used, but some experiment results indicate that $c_1 = c_2 = 1.49$ might provide even better results. Recent work reports that it might be even better to choose a larger cognitive parameter, c_1, than a social parameter, c_2, but with $c_1 + c_2 \leq 4$ [3].

1.3.2 Ant Colony Optimization

Ant Colony Optimization (ACO) deals with artificial systems that is inspired from the foraging behavior of real ants, which are used to solve discrete optimization problems [4]. The main idea is the indirect communication between the ants by means of chemical pheromone trials, which enables them to find short paths between their nest and food.

In nature, ants usually wander randomly, and upon finding food return to their nest while laying down pheromone trails. If other ants find such a path (pheromone trail), they are likely not to keep travelling at random, but to instead follow the trail, returning and reinforcing it if they eventually find food. However, as time passes, the pheromone starts to evaporate. The more time it takes for an ant to travel down the path and back again, the more time the pheromone has to evaporate (and the path to become less prominent). A shorter path, in comparison will be visited by more ants (can be described as a loop of positive feedback) and thus the pheromone density remains high for a longer time.

ACO is implemented as a team of intelligent agents, which simulate the ants behavior, walking around the graph representing the problem to solve using mechanisms of cooperation and adaptation. ACO algorithm requires to define the following: [31, 32]:

- The problem needs to be represented appropriately, which would allow the ants to incrementally update the solutions through the use of a probabilistic transition rules, based on the amount of pheromone in the trail and other problem specific knowledge. It is also important to enforce a strategy to construct only valid solutions corresponding to the problem definition.
- A problem-dependent heuristic function η that measures the quality of components that can be added to the current partial solution.
- A rule set for pheromone updating, which specifies how to modify the pheromone value τ.
- A probabilistic transition rule based on the value of the heuristic function η and the pheromone value τ that is used to iteratively construct a solution.

ACO was first introduced using the Travelling Salesman Problem (TSP). Starting from its start node, an ant iteratively moves from one node to another. When being at a node, an ant chooses to go to a unvisited node at time t with a probability given by

$$p_{i,j}^k(t) = \frac{[\tau_{i,j}(t)]^\alpha [\eta_{i,j}(t)]^\beta}{\sum_{l \in N_i^k} [\tau_{i,j}(t)]^\alpha [\eta_{i,j}(t)]^\beta} \qquad j \in N_i^k \tag{1.20}$$

where N_i^k is the feasible neighborhood of the ant_k, that is, the set of cities which ant_k has not yet visited; $\tau_{i,j}(t)$ is the pheromone value on the edge (i, j) at the time t, α is the weight of pheromone; $\eta_{i,j}(t)$ is a priori available heuristic information on the edge (i, j) at the time t, β is the weight of heuristic information. Two parameters α and β determine the relative influence of pheromone trail and heuristic information. $\tau_{i,j}(t)$ is determined by

$$\tau_{i,j}(t) = \rho \tau_{i,j}(t-1) + \sum_{k=1}^n \Delta \tau_{i,j}^k(t) \qquad \forall(i, j) \tag{1.21}$$

$$\Delta \tau_{i,j}^k(t) = \begin{cases} \frac{Q}{L_k(t)} & \text{if the edge } (i, j) \text{ chosen by the } ant_k \\ 0 & \text{otherwise} \end{cases} \tag{1.22}$$

where ρ is the pheromone trail evaporation rate $(0 < \rho < 1)$, n is the number of ants, Q is a constant for pheromone updating.

More recent work has seen the application of ACO to other problems [33, 34]. A generalized version of the pseudo-code for the ACO algorithm with reference to the TSP is illustrated in Algorithm 3.

Algorithm 3 Ant Colony Optimization Algorithm

01. Initialize the number of ants n, and other parameters.
02. While (the end criterion is not met) do
03. $t = t + 1$;
04.   For $k = 1$ to n
05. ant_k is positioned on a starting node;
06.   For $m = 2$ to $problem\_size$
07.   Choose the state to move into
08.   according to the probabilistic transition rules;
09.   Append the chosen move into $tabu_k(t)$ for the ant_k;
10.   Next m
11.   Compute the length $L_k(t)$ of the tour $T_k(t)$ chosen by the ant_k;
12.   Compute $\Delta\tau_{i,j}(t)$ for every edge (i,j) in $T_k(t)$ according to Eq.(1.22);
13.   Next k
14.   Update the trail pheromone intensity for every edge (i,j) according to Eq.(1.21);
15.   Compare and update the best solution;
16. End While.

1.4 Artificial Neural Networks

How does the brain process information? How is it organized? What are the biological mechanisms involved in brain functioning? These form just a sample of some of the most challenging questions in science. Brains are especially good at performing functions like pattern recognition, motor control, perception, flexible inference, intuition, and guessing. But brains are also slow, imprecise, make erroneous generalizations, are prejudiced, and are usually incapable of explaining their own actions. Neurocomputing, sometimes called brain-like computation or neurocomputation but most often referred to as artificial neural networks (ANN), can be defined as information processing systems (computing devices) designed with inspiration taken from the nervous system, more specifically the brain, and with particular emphasis in problem solving.

Neurocomputing systems are distinct from what is now known as computational neuroscience, which is mainly concerned with the development of biologically-based computational models of the nervous system. Artificial neural networks on the other hand, take a loose inspiration from the nervous system and emphasize the problem solving capability of the systems developed. However, most books on computational neuroscience not only acknowledge the existence of artificial neural networks, but also use several ideas from them in the proposal of more biologically plausible models. They also discuss the artificial neural network suitability as models of real biological nervous systems. Neurons are believed to be the basic units used for computation in the brain, and their simplified abstract models are the basic processing units of neurocomputing devices or artificial neural networks. Neurons are connected to other neurons by a small junction called synapse, whose capability of being modulated is believed to be the basis for most of our cognitive abilities, such as perception, thinking and inferring. Therefore, some essential information about neurons, synapses, and their structural anatomy are relevant for the understanding of how artificial neural networks are designed taking inspiration from biological neural networks. The discussion to be presented here briefly introduces the main aspects of the nervous system used to devise neurocomputing systems, and then focuses on some of the most commonly used artificial neural networks, namely, single- and multi-layer perceptrons, self-organizing networks, and Hopfield networks. The description of the many algorithms uses a matrix notation particularly suitable for their software implementation of the algorithms. The biological plausibility of each model is also assessed. Although neurocomputing can be viewed as a field of research dedicated to the design of intelligent brain-like computers, this report uses the word neurocomputing as a synonym to artificial neural networks [304].

1.4.1 Architecture and Learning Algorithm

Artificial neural networks are computer programs or mathematical representation loosely inspired by the massively connected set of neurons that form the biological neural networks in brain.

The earlier experimental research works about artificial neural networks include:

- McCulloch and Pitts studied the potential and capabilities of the interconnection of several basic components based on the model of a neuron in 1943 [35];
- The name of Perceptron was proposed by Rosenblatt in 1958 [36] ;
- The Perceptron was analyzed, its properties and limitations were given by Minsky and Papert in 1969 [37];
- A number of neural processing models were proposed including the learn matrix in 1960's [38], associative content addressable memory (ACAM) networks [39], and cooperative-competitive neural network models in 1970's [40];
- A particular dynamic network structure was proposed by Hopfield in 1982 [28].

Artificial neural networks are the alternative computing technologies that have been proven useful in a variety of function approximation, pattern recognition, signal processing, system identification and control problems.

The properties and functions of artificial neural networks depend on:

- *The properties of single neuron model.*
  The most common single neuron model is illustrated in Figure 1.7, in which the output of the neuron is the weighted sum of its input x_i,

$u_i = \sum_j \omega_{ij} x_j$, biased by a threshold value θ_i and passed through an activation function f.

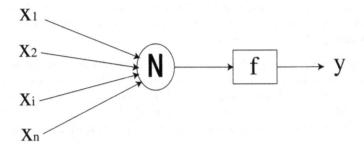

Fig. 1.7 The processing unit of an ANN- a neuron

$$y_i = f\left(\sum_j \omega_{ij} x_j - \theta_i\right) \qquad (1.23)$$

The activation function is selected differently in different applications. Some common choices of selecting activation function in function approximation, system identification and control are shown in Table 1.1.

Table 1.1 The Activation functions

Name	Formula
Sigmoid Function	$f(x) = \frac{1}{1+e^{-x}}$
Gaussian Function	$f(x) = exp(-\frac{x^2}{\sigma^2})$
Symmetric Sigmoid Function	$f(x) = \frac{1-e^{-x}}{1+e^{-x}}$
Hyperbolic Tangent Function	$f(x) = \frac{e^x - e^{-x}}{e^x + e^{-x}}$
Augmented Ratio of Squares	$f(x) = \frac{x^2}{1+x^2} sgn(x)$
Flexible Unipolar Sigmoid Function	$f(x,a) = \frac{2|a|}{1+e^{-2|a|x}}$
Flexible Bipolar Sigmoid Function	$f(x,a) = \frac{1-e^{-2xa}}{a(1+e^{-2xa})}$

- *The topologies of the neural nets* are referred to the number of layers and the ways of the connections of the neurons. Different topologies or architectures of neural nets should have different performance and different computational structure.
- *Parameter tuning techniques* are referred to update the adjustable parameters including the weights, bias, and the parameters used in flexible activation functions. In general, the performance of a certain algorithm

for adaptively tuning the parameters in neural network training stage can be evaluated by its convergence speed, stable properties, robustness and generalization ability. Recently, more and more research works are focused on the stable training problem of artificial neural networks . To some extend, neural network learning problem can be posed as a control problem, therefore, some adaptive control strategies can be introduced directly into the training of artificial neural networks.

Up to date, several kinds of neural network architectures have been developed. Among of them, the multilayer perceptron (MLP), the recurrent neural network (RNN), the fuzzy neural network (FNN), the radial basis function network (RBF) and the wavelet neural network (WNN) are most used neural networks in function approximation, system identification and controller design.

1.4.2   Multilayer Perceptron

Multilayer perceptron is a completely connected feedforward neural network as shown in Figure 1.8. By properly selecting the number of hidden neurons, and the activation function in the hidden layer (i.e., hyperbolic tangent, f) and in the output layer (i.e., flexible sigmoid F), the output of the MLP can be calculated as follows:

$$y_k(t|\theta) = F_k(\sum_{j=0}^{n_h} w_{k,j} h_j(t)) = F_k(\sum_{j=1}^{n_h} w_{k,j} f_j(\sum_{l=0}^{n_i} w_{j,l} x_l(t)) + w_{k,0}) \quad (1.24)$$

where $y_k(t|\theta), (k = 1, 2, \ldots, n_o)$ is the k-th output of neural network, f_j is the j-th activation function for the unit j in the hidden layer and F_k specifies the activation function for output k. $h_j(t)$ is the j-th output of the hidden layer. $w_{k,j}$ and $w_{j,l}$ are the hidden-to-output and input-to-hidden layer weights, respectively. In addition, the bias is regarded as additional weights, i.e., $h_0(t) = x_0(t) = 1$.

1.4.3   Back-Propagation Algorithm

Assume that the used activation functions in the hidden and output layers of the neural network are hyperbolic tangent and flexible bipolar sigmoid function (see Table 1.1).

The derivatives of $F(x, a)$ with respect to the variable x and the parameter a can be obtained as:

$$F'(x, a) = 1 - a^2 F^2(x, a) \quad (1.25)$$

$$F^*(x, a) = \frac{1}{a}[F'(x, a)x - F(x, a)] \quad (1.26)$$

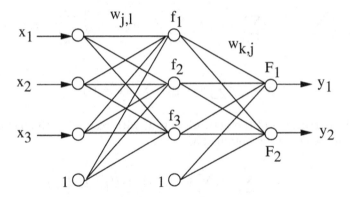

Fig. 1.8 A 3-inputs 2-outputs three-layers perceptron

A batch version of momentum backpropagation algorithm can be easily derived as follows. Given a set of data:

$$Z^N = \{[u(t), y(t)], t = 1, \ldots, N\}$$

Define the objective function as:

$$J = \frac{1}{2N} \sum_{t=1}^{N} \sum_{k=1}^{n_o} (y_k(t) - y_k(t|\theta))^2 = \frac{1}{2N} \sum_{t=1}^{N} \sum_{k=1}^{n_o} \varepsilon_k^2(t, \theta) \qquad (1.27)$$

The general gradient of the least squares criterion takes the form:

$$\frac{\partial J}{\partial \theta} = -\frac{1}{N} \sum_{t=1}^{N} \sum_{k=1}^{n_o} \frac{\partial y_k(t|\theta)}{\partial \theta} [y_k(t) - y_k(t|\theta)] \qquad (1.28)$$

The partial derivatives of the network output with respect to the weights in hidden-to-output layer:

$$\frac{\partial J}{\partial w_{k,j}} = -\frac{1}{N} \sum_{t=1}^{N} h_j(t) F_k'(\sum_{j=0}^{n_h} w_{k,j} h_j(t))(y_k(t) - y_k(t|\theta)) = -\frac{1}{N} \sum_{t=1}^{N} h_j(t) \delta_k(t) \ (1.29)$$

where

$$\delta_k(t) = F_k'(\sum_{j=0}^{n_h} w_{k,j} h_j(t))(y_k(t) - y_k(t|\theta)) \qquad (1.30)$$

Similarly, the partial derivatives of the network output with respect to the weights in input-to-hidden layer cab be obtained as:

$$\frac{\partial J}{\partial w_{j,l}} = -\frac{1}{N} \sum_{t=1}^{N} x_l(t) f_j'(\sum_{l=0}^{n_i} w_{j,l} x_l(t)) \sum_{k=1}^{n_o} w_{k,j} \delta_k(t) = -\frac{1}{N} \sum_{t=1}^{N} x_l(t) \delta_j(t) \quad (1.31)$$

where

$$\delta_j(t) = f_j'(\sum_{l=0}^{n_i} w_{j,l} x_l(t)) \sum_{k=1}^{n_o} w_{k,j} \delta_k(t) \quad (1.32)$$

The partial derivatives of the network output with respect to the parameter a_k in output layer cab be obtained as:

$$\frac{\partial J}{\partial a_k} = -\frac{1}{N} \sum_{t=1}^{N} F_k^*(\sum_{j=0}^{n_h} w_{k,j} h_j(t))(y_k(t) - y_k(t|\theta)) \quad (1.33)$$

Therefore, the weights in hidden-to-output layer and in input-to-hidden layer, and parameters in output layer can be updated by:

$$w_{k,j} = w_{k,j} + \eta_1 \frac{\partial J}{\partial w_{k,j}} + \alpha_1 \Delta w_{k,j} \quad (1.34)$$

$$w_{j,l} = w_{j,l} + \eta_2 \frac{\partial J}{\partial w_{j,l}} + \alpha_2 \Delta w_{j,l} \quad (1.35)$$

$$a_k = a_k + \eta_3 \frac{\partial J}{\partial a_k} + \alpha_3 \Delta a_k \quad (1.36)$$

1.4.4　Evolutionary Algorithm Based Training

The strong interest in neural networks in the scientific community is fueled by the many successful and promising applications especially to tasks of optimization, speech recognition, pattern recognition, signal processing, function approximation, control problems , financial modeling etc.. Even though artificial neural networks are capable of performing a wide variety of tasks, yet in practice sometimes they deliver only marginal performance. Inappropriate topology selection and learning algorithm are frequently blamed. There is little reason to expect that one can find a uniformly best algorithm for selecting the weights in a feedforward artificial neural network. This is in accordance with the no free lunch theorem, which explains that for any algorithm, any elevated performance over one class of problems is exactly paid for in performance over another class [292].

At present, neural network design relies heavily on human experts who have sufficient knowledge about the different aspects of the network and the problem domain. As the complexity of the problem domain increases, manual design becomes more difficult and unmanageable. Some global optimization techniques like evolutionary programming, simulated annealing and genetic algorithms have also been used for the training of ANNs. Reader may consult [41] for a nice survey of such research.

Evolutionary design of artificial neural networks has been widely explored [50]. Evolutionary algorithms are used to adapt the connection weights, network architecture and learning rules according to the problem environment. A distinct feature of evolutionary neural networks is their adaptability to a dynamic environment. In other words, such neural networks can adapt to an environment as well as changes in the environment. The two forms of adaptation: evolution and learning in evolutionary artificial neural networks make their adaptation to a dynamic environment much more effective and efficient than the conventional learning approach [293] [295] [294]. It has been shown that the binary coding scheme used in genetic algorithm is neither necessary nor beneficial [43], [44]. In addition, Fogel and Ghozeil [44] illustrated that under some fairly general assumptions, there are fundamental equivalences between various representations. Several successful studies using real values instead of binary coding scheme include Montana and Davis [45] and Sexton et al., [46] , [49] for GA and Porto et al., [47] and Saravanan et al., [48] for other evolutionary algorithms.

The performance of evolutionary design is very often described by effectiveness, ease of use and efficiency. *Effectiveness* refers to the accuracy of each algorithm to estimate the true functional form. *Ease-of-use* deals with the effort needed for optimal algorithm settings for the problems at hand. *Efficiency* of an algorithm is computed by comparing the CPU time needed for converging upon the best found solutions.

1.4.5 Self Organizing Feature Maps

Self Organizing Feature Maps (SOFM) are a data visualization technique proposed by Professor Teuvo Kohonen [300], which reduce the dimensions of data through the use of self-organizing neural networks. A SOFM learns the categorization, topology and distribution of input vectors. SOFM allocate more neurons to recognize parts of the input space where many input vectors occur and allocate fewer neurons to parts of the input space where few input vectors occur. Neurons next to each other in the network learn to respond to similar vectors. SOFM can learn to detect regularities and correlations in their input and adapt their future responses to that input accordingly. An important feature of SOFM learning algorithm is that it allow neurons that are neighbors to the winning neuron to output values. Thus the transition of output vectors is much smoother than that obtained with competitive layers, where only one neuron has an output at a time. The problem that data visualization attempts to solve is that humans simply cannot visualize high dimensional data. The way SOFM go about reducing dimensions is by producing a map of usually 1 or 2 dimensions, which plot the similarities of the data by grouping similar data items together (data clustering). In this process SOFM accomplish two things, they reduce dimensions and display similarities. It is important to note that while a self-organizing map does not take long to organize itself so that neighboring neurons recognize similar inputs, it can take a long time for the map to finally arrange itself according to the distribution of input vectors.

1.4.6 Radial Basis Function

The Radial Basis Function (RBF) network is a three-layer feed-forward network that uses a linear transfer function for the output units and a nonlinear transfer function (normally the Gaussian) for the hidden layer neurons [299]. Radial basis networks may require more neurons than standard feed-forward backpropagation networks, but often they can be designed with lesser time. They perform well when many training data are available. Much of the inspiration for RBF networks has come from traditional statistical pattern classification techniques. The input layer is simply a fan-out layer and does no processing. The second or hidden layer performs a non-linear mapping from the input space into a (usually) higher dimensional space whose activation function is selected from a class of functions called basis functions. The final layer performs a simple weighted sum with a linear output. Contrary to backpropagation networks the weights of the hidden layer basis units (input to hidden layer) are set using some clustering techniques. The idea is that the patterns in the input space form clusters. If the centers of these clusters are known, then the Euclidean distance from the cluster center can be measured. As the input data moves away from the connection weights, the activation value reduces. This distance measure is made non-linear in such a way that for input data close to a cluster center gets a value close to 1. Once the hidden layer weights are set, a second phase of training (usually backpropagation) is used to adjust the output weights.

1.4.7 Recurrent Neural Networks

Recurrent networks are the state-of-the-art in nonlinear time series prediction, system identification, and temporal pattern classification. As the output of the network at time t is used along with a new input to compute the output of the network at time t+1, the response of the network is dynamic [296]. Time Lag Recurrent Networks (TLRN) are multi-layered perceptrons extended with short-term memory structures that have local recurrent connections. The recurrent neural network is a very appropriate model for processing temporal (time-varying) information. Examples of temporal problems include time series prediction, system identification and temporal pattern recognition. A simple recurrent neural network could be constructed by a modification of the multi-layered feed-forward network with the addition of a context layer. The context layer is added to the structure, which retains information between observations. At each time step, new inputs are fed to the network. The previous contents of the hidden layer are passed into the context layer. These then feed back into the hidden layer in the next time step. Initially, the context layer contains nothing, so the output from the hidden layer after the first input to the network will be the same as if there is no context layer. Weights are calculated in the same way for the new connections from and to the context layer from the hidden layer. The training algorithm used in TLRN (backpropagation through time) is more advanced than standard backpropagation algorithm. Very often TLRN requires a smaller network to learn temporal problems when compared to MLP that use extra inputs to represent the past samples. TLRN is biologically more plausible and computationally more powerful than other adaptive models such as the hidden Markov model. Some popular recurrent network architectures are the Elman recurrent network in which the hidden unit activation values are fed back to an extra set of input units and the Jordan recurrent network in which output values are fed back into hidden units.

1.4.8 Adaptive Resonance Theory

Adaptive Resonance Theory (ART) was initially introduced by Professor Stephen Grossberg [301] as a theory of human information processing. ART neural networks are extensively used for supervised and unsupervised classification tasks and function approximation. There are many different variations of ART networks available today [297]. For example, ART1 performs unsupervised learning for binary input patterns, ART2 is modified to handle both analog and binary input patterns, and ART3 performs parallel searches of distributed recognition codes in a multilevel network hierarchy. Fuzzy ARTMAP represents a synthesis of elements from neural networks, expert systems, and fuzzy logic.

1.5 Fuzzy Systems

Fuzzy sets were introduced by Professor Lotfi Zadeh [302] with a view to reconcile mathematical modeling and human knowledge in the engineering sciences. Since then, a considerable body of literature has blossomed around the concept of fuzzy sets in an incredibly wide range of areas, from mathematics and logics to traditional and advanced engineering methodologies.

1.5.1 The Definition of Fuzzy Sets

To introduce fuzzy sets consider the $X = \{x_1, x_2, x_3, x_4, x_5\}$ crisp set that will be called universe, or universal set and let $Y \subset x = \{x_1, x_2, x_3\}$ is its crisp subset.

By using the characteristic function defined as:

$$\mu_Y(x) = \begin{cases} 1, \; if \; x \in Y \\ 0, \; otherwise \end{cases} \quad (1.37)$$

The subset Y can be uniquely represented by ordered pairs:

$$Y = \{(x_1, 1), (x_2, 1), (x_3, 0), (x_4, 0), (x_5, 1)\} \quad (1.38)$$

Zadeh proposed that the second member of an ordered pair (which is called the membership grade of the appropriate element) can take its value not only from the set $\{0, 1\}$ but from the closed interval $[0, 1]$ as well. By using this idea fuzzy sets are defined as follows:

Definition. Let X a universal crisp set. The set of ordered pairs

$$Y = \{(x, \mu_Y(x))|x \in X, \mu_Y : X \to [0, 1]\} \quad (1.39)$$

is said to be the fuzzy subset of X. The $\mu_Y : X \to [0, 1]$ function is called as membership function and its vlaue is said to be the membership grade of x.

1.6   Takagi-Sugeno Fuzzy Model

The world of information is surrounded by uncertainty and imprecision. The human reasoning process can handle inexact, uncertain and vague concepts in an appropriate manner. Usually, the human thinking, reasoning and perception process cannot be expressed precisely. These types of experiences can rarely be expressed or measured using statistical or probability theory. Fuzzy logic provides a framework to model uncertainty, human way of thinking, reasoning and the perception process [305]. A fuzzy expert system is simply an expert system that uses a collection of fuzzy membership functions and rules, instead of Boolean logic, to reason about data. The rules in a fuzzy expert system are usually of a form similar to the following: If A is low and B is high then X = medium where A and B are input variables, X is an output variable. Here low, high, medium are fuzzy sets defined on A, B and X respectively. The antecedent (the rule's premise) describes to what degree the rule applies, while the the rule's consequent assigns a membership function to each of one or more output variables.

A fuzzy model proposed by Takagi and Sugeno [51] is described by fuzzy if-then rules whose consequent parts are represented by linear equations. This fuzzy model is of the following form:

$$R_i : If \; x_1 \; is \; A_{i1} \ldots, x_n \; is \; A_{in} \; then \; y_i = c_{i0} + c_{i1}x_1 + \cdots + c_{in}x_n \quad (1.40)$$

where $i = 1, 2, \ldots, N$, N is the number of if-then rules, $c_{ik}(k = 0, 1, \ldots, n)$ are the consequent parameters, y_i is the output from the ith if-then rule, and A_{ik} is a fuzzy set.

Given an input $(x_1, x_2, \ldots, x_n)$, the final output of the fuzzy model is referred as follows:

$$y = \frac{\sum_{i=1}^{N} \omega_i y_i}{\sum_{i=1}^{N} \omega_i} = \frac{\sum_{i=1}^{N} \omega_i (c_{i0} + c_{i1} x_1 + \cdots + c_{in} x_n)}{\sum_{i=1}^{N} \omega_i} = \frac{\sum_{k=0}^{n} \sum_{i=1}^{N} \omega_i c_{ik} x_k}{\sum_{i=1}^{N} \omega_i} \quad (1.41)$$

where $x_0 = 1$, ω_i is the weight of the ith IF-THEN rule for the input and is calculated as:

$$\omega_i = \prod_{k=1}^{n} A_{ik}(x_k), \quad (1.42)$$

where $A_{ik}(x_k)$ is the grad of membership of x_k in A_{ik}.

1.6.1 Universal Approximation Property

To Takagi-Sugeno approach, the universal approximation property was proved in [52, 53]. In addition, a natural further generalization of this approach was proposed in [54, 55], in which in the conclusion of each rule, the desired output y is given not by an explicit formula, but by a (crisp) dynamical systems, i.e., by a system of differential equations that determine the time derivative of the output variable (i.e., its change in time) as a function of the inputs and of the previous values of output. This generalization also has universal approximation property.

A simplified Takagi-Sugeno fuzzy model proposed by Hao [57] has the following rule base:

$$R_i : If \ x_1 \ is \ A_{i1} \ \ldots, x_n \ is \ A_{in} \ then \ y_i = k_i (c_0 + c_1 x_1 + \cdots + c_n x_n) \quad (1.43)$$

where $i = 1, 2, \ldots, N$, N is the number of if-then rules. From this it can be seen that the free parameters in the consequent part of the IF-THEN rules are reduced significantly. The universal approximation property of this simplified T-S fuzzy model has also been proved, and successfully applied to the identification and control of nonlinear systems.

1.6.2 Fuzzy Expert Systems - Design Challenges

Fuzzy logic systems have been successfully applied to a vast number of scientific and engineering problems in recent years. The advantage of solving the complex nonlinear problems by utilizing fuzzy logic methodologies is that the experience or expert's knowledge described as a fuzzy rule base can be

directly embedded into the systems for dealing with the problems. A number
of improvements have been made in the aspects of enhancing the systematic
design method of fuzzy logic systems.

Expert knowledge is often the main source to design the fuzzy expert systems. According to the performance measure of the problem environment,
the membership functions, rule bases and the inference mechanism are to be
adapted [307]. Neural network learning, self-organizing maps and clustering
methods could be used to generate rules. Gradient descent and its variants
could be applied to fine-tune the parameters of parameterized input/output
membership functions and fuzzy operators [308], [310]. Adaptation of fuzzy
inference systems using evolutionary computation techniques has been widely
explored. Automatic adaptation of membership functions is popularly known
as self tuning and the chromosome encodes parameters of trapezoidal, trian-
gle, logistic, hyperbolic-tangent, Gaussian membership functions etc. Evolutionary search of fuzzy rules can be carried out using three approaches. In
the first method (Michigan approach) the fuzzy knowledge base is adapted as
a result of antagonistic roles of competition and cooperation of fuzzy rules.
The second method (Pittsburgh approach) evolves a population of knowledge bases rather than individual fuzzy rules. Reproduction operators serve
to provide a new combination of rules and new rules. The third method (iterative rule learning approach) is very much similar to the first method with
each chromosome representing a single rule, but contrary to the Michigan
approach, only the best individual is considered to form part of the solution,
discarding the remaining chromosomes in the population. The evolutionary
learning process builds up the complete rule base through a iterative learning
process [311].

There are lots of challenges and remaining problems to be solved. For example, how to automatically partition the input space for each input-output
variables, how many fuzzy rules are really needed for properly approximating
the unknown nonlinear systems, and how to determine it automatically and
so on. As is well known, the curse-of-dimensionality is an unsolved problem
in the field.

1.7 Probabilistic Computing

Probabilistic models are viewed as similar to that of a game, actions are
based on expected outcomes. The center of interest moves from the deterministic to probabilistic models using statistical estimations and predictions.
In the probabilistic modeling process, risk means uncertainty for which the
probability distribution is known. Therefore risk assessment means a study to
determine the outcomes of decisions along with their probabilities. Decision-
makers often face a severe lack of information. Probability assessment quan-
tifies the information gap between what is known, and what needs to be
known for an optimal decision. The probabilistic models are used for protec-
tion against adverse uncertainty, and exploitation of propitious uncertainty

[313]. A good example is the probabilistic neural network (Bayesian learning) in which probability is used to represent uncertainty about the relationship being learned. Before we have seen any data, our prior opinions about what the true relationship might be can be expressed in a probability distribution over the network weights that define this relationship. After we look at the data, our revised opinions are captured by a posterior distribution over network weights. Network weights that seemed plausible before, but which don't match the data very well, will now be seen as being much less likely, while the probability for values of the weights that do fit the data well will have increased. Typically, the purpose of training is to make predictions for future cases in which only the inputs to the network are known. The result of conventional network training is a single set of weights that can be used to make such predictions.

1.8 Hybrid Intelligent Systems

Several adaptive hybrid intelligent systems have in recent years been developed for model expertise, image and video segmentation techniques, process control, mechatronics, robotics and complicated automation tasks etc. Many of these approaches use the combination of different knowledge representation schemes, decision making models and learning strategies to solve a computational task. This integration aims at overcoming limitations of individual techniques through hybridization or fusion of various techniques. These ideas have led to the emergence of several different kinds of intelligent system architectures. It is well known that the intelligent systems, which can provide human like expertise such as domain knowledge, uncertain reasoning, and adaptation to a noisy and time varying environment, are important in tackling practical computing problems. Most of the current Hybrid Intelligent Systems (HIS) consists of 3 essential paradigms: artificial neural networks, fuzzy inference systems and global optimization algorithms (example, evolutionary algorithms). Nevertheless, HIS is an open instead of conservative concept. That is, it is evolving those relevant techniques together with the important advances in other new computing methods. To realize a highly intelligent system, a synthesis of various techniques is required. Each technique plays a very important role in the development of HIS. Experience has shown that it is crucial for the design of HIS to primarily focus on the integration and interaction of different techniques rather than merge different methods to create ever-new techniques. Techniques already well understood, should be applied to solve specific domain problems within the system. Their weakness must be addressed by combining them with complementary methods. Due to the complementary features and strengths of different systems, the trend in the design of hybrid system is to merge both of them into a more powerful integrated system, to overcome their individual weakness (example: global optimization algorithms could be useful to formulate an optimal combination of neural networks and fuzzy inference systems) .

1.9   Models of Hybrid Intelligent Systems

The various HIS architectures could be broadly classified into 3 different categories based on the systems overall architecture: (1) Transformational Architectures (2) Hierarchical Hybrid Architectures and (3) Integrated Hybrid Architectures. The following sections discuss each of these strategies and expected uses of the model, and benefits and limitations of the approach.

In a transformational hybrid model the system begins as one type of system and end up as the other. Determining which technique is used for development and which is used for delivery is based on the desirable features that the technique offers. Expert Systems and ANNs have proven to be useful transformational models. Variously, either the expert system is incapable of adequately solving the problem, or the speed, adaptability, and robustness of neural network is required. Knowledge from the expert system is used to set the initial conditions and training set for ANN. Transformational hybrid models are often quick to develop and ultimately require maintenance on only one system. Most of the developed models are just application oriented.

The architecture is built in a hierarchical fashion, associating a different functionality with each layer. The overall functioning of the model will depend on the correct functioning of all the layers. Possible error in one of the layers will directly affect the desired output.

Fused architectures are the first true form of integrated intelligent systems. These models include systems, which combine different techniques into one single computational model. They share data structures and knowledge representations. Another approach is to put the various techniques on a side-by-side basis and focus on their interaction in the problem-solving task. This method might allow integrating alternative techniques and exploiting their mutuality. The benefits of fused architecture include robustness, improved performance and increased problem-solving capabilities. Finally, fully integrated models can provide a full range of capabilities such as adaptation, generalization, noise tolerance and justification. Fused systems have limitations caused by the increased complexity of the inter module interactions and specifying, designing, and building fully integrated models is complex.

Part II

Flexible Neural Trees

Flexible Neural Tree: Foundations and Applications

2.1 Introduction to Flexible Neural Tree

Artificial neural networks (ANNs) have been successfully applied to a number of scientific and engineering fields in recent years, i.e., function approximation, system identification and control, image processing, time series prediction [58]. A neural network's performance is highly dependent on its structure. The interaction allowed between the various nodes of the network is specified using the structure only. An artificial neural network structure is not unique for a given problem, and there may exist different ways to define a structure corresponding to the problem. Depending on the problem, it may be appropriate to have more than one hidden layer, feedforward or feedback connections, or in some cases, direct connections between input and output layer.

There has been a number of attempts for designing ANN architectures automatically. The early methods include constructive and pruning algorithms [59–61]. The main problem of these methods is that the topological subsets rather than the complete class of artificial neural network architecture is searched in the search space by structural hill climbing methods [62]. Recent tendencies to optimize artificial neural network architecture and weights include EPNet [63][64] and the NeuroEvolution of Augmenting Topologies (NEAT) [65]. Utilizing a tree to represent a NN-like model is motivated by the work of Zhang, where a method of evolutionary induction of the sparse neural trees was proposed [66]. Based on the representation of neural tree, architecture and weights of higher order sigma-pi neural networks were evolved by using genetic programming and breeder genetic algorithm, respectively.

In this Chapter, we illustrate a general and enhanced flexible neural tree (FNT) models for problem solving. Based on the pre-defined instruction/operator sets, a flexible neural tree model can be created and evolved. In this approach, over-layer connections, different activation functions for different nodes and input variables selection are allowed. The hierarchical structure could be evolved by using tree-structure based evolutionary algorithms with specific instructions. The fine tuning of the parameters encoded in the

Y. Chen, A. Abraham.: Tree-Struc. Based Hybrid Com. Intelligence, ISRL 2, pp. 39–96.
springerlink.com   © Springer-Verlag Berlin Heidelberg 2010

structure could be accomplished by using parameter optimization algorithms. The flexible neural tree method interleaves both optimizations. Starting with random structures and corresponding parameters, it first tries to improve the structure and then as soon as an improved structure is found, it fine tunes its parameters. It then goes back to improving the structure again and, provided it finds a better structure, it again fine tunes the rules' parameters. This loop continues until a satisfactory solution is found or a time limit is reached.

This chapter provides a detailed introduction to the flexible neural tree algorithm development and is first illustrated in some function approximation problems and also in some real world problems like intrusion detection, exchange rate forecasting, face recognition, cancer detection and protein fold recognition. Further the multi-input multi-output flexible neural trees algorithm is introduced and is illustrated for some problem solving. Finally an ensemble of flexible neural tree is introduced and is illustrated for stock market prediction problem.

2.2 Flexible Neural Tree Algorithms

2.2.1 Encoding and Evaluation

A tree-structural based encoding method with specific instruction set is selected for representing a flexible neural tree model. The reason for choosing the representation is that the tree can be created and evolved using the existing or modified tree-structure-based approaches, i.e., Genetic Programming (GP), Probabilistic Incremental Program Evolution (PIPE) and Ant Programming (AP).

2.2.2 Flexible Neuron Instructor

The used function set F and terminal instruction set T for generating a flexible neural tree model are described as follows:

$$S = F \bigcup T = \{+_2, +_3, \ldots, +_N\} \bigcup \{x_1, \ldots, x_n\}, \qquad (2.1)$$

where $+_i(i = 2, 3, \ldots, N)$ denote non-leaf nodes' instructions and taking i arguments. $x_1, x_2, \ldots, x_n$ are leaf nodes' instructions and taking no argument each. The output of a non-leaf node is calculated as a flexible neuron model (see Figure 2.1). From this point of view, the instruction $+_i$ is also called a flexible neuron operator with i inputs.

In the construction process of a neural tree, if a nonterminal instruction, i.e., $+_i(i = 2, 3, 4, \ldots, N)$ is selected, i real values are randomly generated and used for representing the connection strength between the node $+_i$ and its children. In addition, two adjustable parameters a_i and b_i are randomly created as flexible activation function parameters. The flexible activation function used is given by:

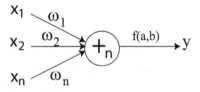

Fig. 2.1 A flexible neuron operator

$$f(a_i, b_i, x) = e^{-(\frac{x-a_i}{b_i})^2}. \tag{2.2}$$

The output of a flexible neuron $+_n$ is calculated as follows and the total excitation of $+_n$ is given by:

$$net_n = \sum_{j=1}^{n} w_j * x_j \tag{2.3}$$

where $x_j (j = 1, 2, \ldots, n)$ are the inputs to node $+_n$. The output of the node $+_n$ is then calculated by:

$$out_n = f(a_n, b_n, net_n) = e^{-(\frac{net_n - a_n}{b_n})^2}. \tag{2.4}$$

A typical flexible neural tree model is illustrated in Figure 2.2. The overall output of flexible neural tree can be computed from left to right by depth-first method, recursively.

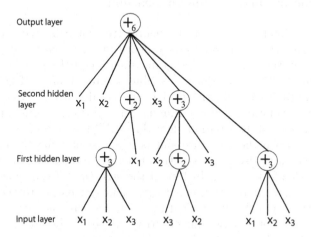

Fig. 2.2 A typical representation of neural tree with function instruction set $F = \{+_2, +_3, +_4, +_5, +_6\}$, and terminal instruction set $T = \{x_1, x_2, x_3\}$

2.2.3   Fitness Function

A fitness function maps flexible neural tree to scalar, real-valued fitness values that reflect the flexible neural tree's performances on a given task. Firstly the fitness functions should be seen as error measures, i.e., MSE or $RMSE$. A secondary non-user-defined objective for which algorithm always optimizes flexible neural tree is FNT size as measured by number of nodes. Among flexible neural tree's with equal fitness smaller ones are always preferred. The fitness function used for PIPE and Simulated Annealing (SA) is given by mean square error (MSE):

$$Fit(i) = \frac{1}{P} \sum_{j=1}^{P} (y_1^j - y_2^j)^2 \tag{2.5}$$

or Root Mean Square Error ($RMSE$):

$$Fit(i) = \sqrt{\frac{1}{P} \sum_{j=1}^{P} (y_1^j - y_2^j)^2} \tag{2.6}$$

where P is the total number of samples, y_1^j and y_2^j are the actual time-series and the flexible neural tree model output of j-th sample. $Fit(i)$ denotes the fitness value of i-th individual.

2.2.4   Structure and Parameter Learning

Finding an optimal or near-optimal neural tree could be accomplished by using tree-structure based evolutionary algorithms, i.e., genetic programming (GP), probabilistic incremental program evolution (PIPE), gene expression programming (GEP), estimation of distribution programming (EDP) and the parameters optimization algorithms, i.e., genetic algorithms (GA), evolution strategy (ES), evolutionary programming (EP), particle swarm optimization (PSO), estimation of distribution algorithm (EDA), and so on.

In order to learn the structure and parameters of a flexible neural tree simultaneously, a tradeoff between the structure optimization and parameter learning should be taken. In fact, if the structure of the evolved model is not appropriate, it is not useful to pay much attention to the parameter optimization. On the contrary, if the best structure has been found, the further structure optimization may destroy the best structure. We illustrate a technique for balancing the structure optimization and parameter learning. If a better structure is found then do local search for a number of steps:

maximum allowed steps or stop in case no better parameter vector is found
for a significantly long time (say 100 to 2000 in our experiments). Where the
criterion of better structure is distinguished as follows: if the fitness value of
the best program is smaller than the fitness value of the elitist program, or
the fitness values of two programs are equal but the nodes of the former is
lower than the later, then we say that the better structure is found.

To find the optimal parameters set (weights and activation function parameters) of a flexible neural tree model, there are a number of global and
local search algorithms, i.e., genetic algorithm, evolutionary programming,
gradient based learning method etc. A variant of simulated annealing (called
degraded ceiling) is selected due to its straightforward property and fast local
search capability.

Simulated annealing (SA) is one of the most widely studied local search
meta-heuristics. It was proposed as a general stochastic optimization technique and has been applied to solve a wide range of problems.

The basic ideas of the simulated annealing search are that it accepts worse
solutions with a probability $p = e^{-\frac{\delta}{T}}$, where $\delta = f(s^*) - f(s)$, the s and
s^* are the old and new solution vectors, $f(s)$ denotes the cost function, the
parameter T denotes the temperature in the process of annealing. Originally
it was suggested to start the search from a high temperature and reduce it to
the end of the process by a formula: $T_{i+1} = T_i - T_i * \beta$. However, the cooling
rate β and initial value of T should be carefully selected due to it is problem
dependent.

Algorithm 4 Flexible Neural Tree (FNT): General Learning Algorithm

---

01. Set the initial values of parameters used in the PIPE and SA algorithms. Set
the elitist program as NULL and its fitness value as a biggest positive real number
of the computer at hand. Create the initial population (flexible neural trees and
their corresponding parameters)

02. Structure optimization by PIPE algorithm as described in subsection 3.1, in
which the fitness function is calculated by mean square error (MSE) or root mean
square error(RMSE)

03. If the better structure is found, then go to step 4), otherwise go to step 2)

04.Parameter optimization by degraded ceiling algorithm as described in subsection
3.2. In this stage, the tree structure or architecture of flexible neural tree model
is fixed, and it is the best tree taken from the end of run of the PIPE search.
All of the parameters used in the best tree formulated a parameter vector to be
optimized by local search

05.If the maximum number of local search is reached, or no better parameter vector
is found for a significantly long time (100 steps) then go to step 6); otherwise go
to step 4)

06.If satisfied solution is found, then stop; otherwise go to step 2)

2.2.5   Flexible Neural Tree Applications

Function Approximation

A non-linear static benchmark modelling problem [68, 95] is considered for illustration, which is described by:

$$y = (1 + x_1^{-2} + x_2^{-1.5})^2, \quad 1 \le x_1, x_2 \le 5. \tag{2.7}$$

50 training and 200 test samples are randomly generated within the interval $[1, 5]$. The static nonlinear function is approximated by using the neural tree model with the pre-defined instruction sets $I = \{+_2, +_3, \ldots, +_8, x_0, x_1\}$, where x_0 and x_1 denote input variables x_1 and x_2 of the static nonlinear function respectively. The initial parameters used in PIPE are depicted in Table 2.1.

The evolved neural tree model is obtained at generation 15 with MSE value 0.000737 for training data set and 0.00086 for validation data set, respectively. The optimized neural tree is illustrated in Figure 2.3. The optimal weights are w_1 to w_{30} are 2.09, -0.01, 3.01, 4.16, 0.72, 4.16, -0.49, 1.15, 1.96, 2.31, 1.58, 2.81, 0.08, 2.12, 1.13, 4.26, 0.65, 0.14, -0.33, -0.86, 0.69, 0.53, 0.28, -0.02, 0.71, 3.13, 0.38, -0.36, -0.97, -0.93, respectively. The optimized activation functions from f_1 to f_{12} are listed in Table 2.2.

Figures 2.4 and 2.5 present the outputs of actual static nonlinear function and the evolved neural tree model, and the prediction errors for training data set and validation data set, respectively. Table 2.3 contains comparison results of different models for the static function approximation and provides, in

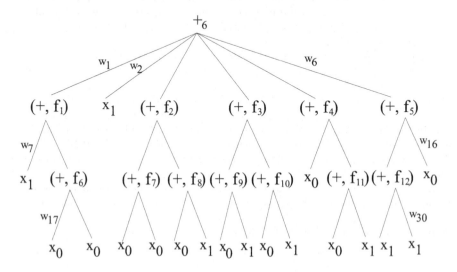

Fig. 2.3 Evolved neural tree for approximating a static nonlinear function

Table 2.1 Parameters used in the PIPE algorithm for architecture optimization of the neural tree

Population size PS	100
Elitist learning probability P_{el}	0.01
Learning rate lr	0.01
Fitness constant ε	0.000001
Overall mutation probability P_M	0.4
Mutation rate mr	0.4
Prune threshold T_P	0.999
Maximum random search steps	2000
Initial connection weights	rand[-1, 1]
Initial parameters a_p and b_p	rand[0,1]

Table 2.2 The optimized flexible activation function parameters for static function approximation

f_1		f_2		f_3		f_4	
a	b	a	b	a	b	a	b
2.74	2.47	3.18	3.17	4.76	2.48	0.08	2.12
f_5		f_6		f_7		f_8	
a	b	a	b	a	b	a	b
2.26	2.40	1.88	2.59	1.67	2.15	0.74	0.93
f_9		f_{10}		f_{11}		f_{12}	
a	b	a	b	a	b	a	b
0.27	0.46	-0.10	0.58	2.67	2.57	1.70	2.42

addition, results achieved with the neural tree model developed in this paper. It is obvious that the flexible neural tree model worked well for generating an approximating model of the static non-linear system.

Nonlinear Systems Identification

A second-order non-minimum phase system with gain 1, time constants $4s$ and $10s$, a zero at $1/4s$, and output feedback with a parabolic nonlinearity is chosen to be identified [75]. With sampling time $T_0 = 1s$ this system follows the nonlinear difference equation

$$y(k) = -0.07289[u(k-1) - 0.2y^2(k-1)] + 0.09394[u(k-2) - 0.2y^2(k-2)]$$

$$+ 1.68364y(k-1) - 0.70469y(k-2). \tag{2.8}$$

where the input lie in the interval [-1,1].

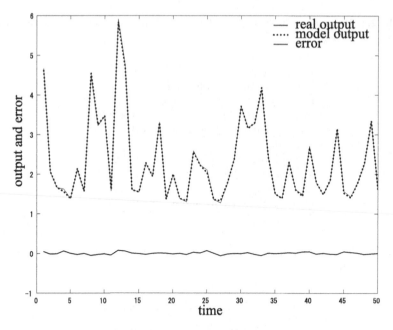

Fig. 2.4 Outputs of actual system, the neural tree model and the approximation error: for training data samples

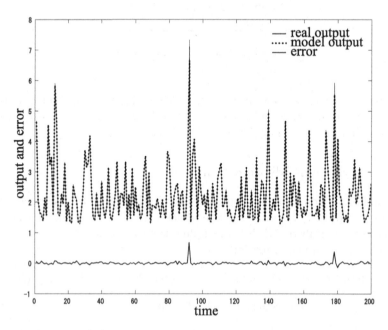

Fig. 2.5 Outputs of actual system, the neural tree model and the approximation error: for test samples

Table 2.3 Comparison of models for the non-linear function approximation

Author	Learning method	MSE
Lin [71]	Back-propagation	0.005
Sugeno [69]	Fuzzy+WLS	0.01
Delgado [72]	Fuzzy Clustering	0.231
Russo [70]	GA+NN+fuzzy	0.00078
Kukolj [73]	Fuzzy clustering+LS	0.0015
FNT	Neural Tree	0.00086

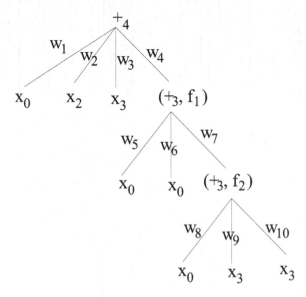

Fig. 2.6 Evolved neural tree for identification of a second order non-minimum phase system

A training and a test sequence of 1000 samples each were generated. The input sequence for training consists of pulses with random amplitude in the range [-1, 1] and with random duration between 1 and 10 sampling periods (Figure 2.7(upper)). The input sequence for test consists of specific pulses shown in Figure 2.7 (lower).

The used instruction sets are $I_0 = \{+_2, +_3, \ldots, +_{10}\}$, $I_1 = \{+_2, +_3, u(k-1), u(k-2), y(k-1), y(k-2)\}$ and $I_2 = \{x_0, x_1, x_2, x_3\}$. Where x_0, x_1, x_2 and x_3 represents $u(k-1)$, $u(k-2)$, $y(k-1)$ and $y(k-2)$, respectively. The output is $y(k)$. The parameters used in PIPE evolution process are also the same as those described in Table 2.2.

After 5 generations of the evolution, the optimal neural tree model was obtained with the MSE value 0.000394. The MSE value for test data set is 0.000437. It is clear that Both the training and test error of the neural tree

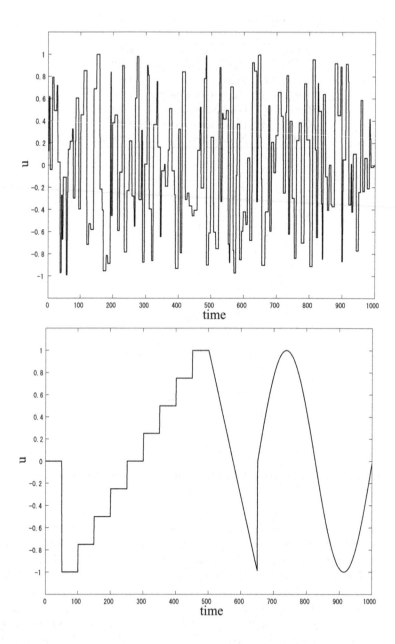

Fig. 2.7 The input signals for generating the excitation and test data sets of the dynamic system. The input signal for creating the training data set (upper) and the for the test data set (lower).

model are smaller than the errors of the local linear neural-fuzzy model [75]. The evolved optimal weights from w_1 to w_{10} are 0.05, -0.39, -0.77, 1.72, 0.48, 0.86, 0.51, 0.49, -0.18 and -0.74, respectively. A comparison has been made to show the actual output, the neural tree model output and the identification error of the dynamic system. (see Figure 2.8).

From above simulation results, it is evident that the proposed neural tree model works very well for nonlinear function approximation, time series prediction and dynamic system identification problems.

Intrusion Detection

The traditional prevention techniques such as user authentication, data encryption, avoiding programming errors and firewalls are used as the first line of defense for computer security. Recently, Intrusion Detection Systems (IDS) have been used in monitoring attempts to break security, which provides important information for timely countermeasures. Intrusion detection is classified into two types: misuse intrusion detection and anomaly intrusion detection. Misuse intrusion detection uses well-defined patterns of the attack that exploit weaknesses in system and application software to identify the intrusions. Anomaly intrusion detection identifies deviations from the normal usage behavior patterns to identify the intrusion.

Data mining approaches for intrusion detection was first implemented in Mining Audit Data for Automated Models for Intrusion Detection [79],[78]. Raw data is converted into ASCII network packet information, which in turn is converted into connection level information. These connection level records contain within connection features like service, duration etc. Data mining algorithms are applied to this data to create models to detect intrusions. Neural networks have been used both in anomaly intrusion detection as well as in misuse intrusion detection. In the first approach of neural networks [80] for intrusion detection, the system learns to predict the next command based on a sequence of previous commands by a user. Support Vector Machines (SVM) have proven to be a good candidate for intrusion detection because of its training speed and scalability. Besides SVMs are relatively insensitive to the number of data points and the classification complexity does not depend on the dimensionality of the feature space, so they can potentially learn a larger set of patterns and scale better than neural networks [82]. Neuro-fuzzy computing is a popular framework for solving complex problems. An Adaptive neuro-fuzzy IDS is proposed by Shah et al. [85]. Multivariate Adaptive Regression Splines (MARS) is an innovative approach that automates the building of accurate predictive models for continuous and binary dependent variables [81]. It excels at finding optimal variable transformations and interactions, and the complex data structure that often hides in high-dimensional data. An IDS based on MARS technology is proposed in [84]. Linear Genetic Programming (LGP) is a variant of the conventional Genetic Programming (GP) technique that acts on linear genomes. Its main characteristics in

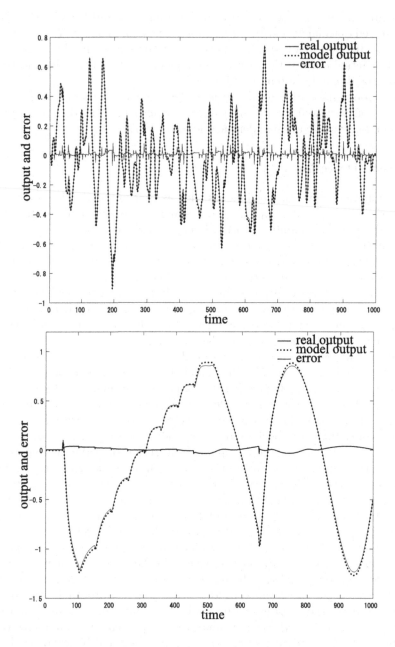

Fig. 2.8 Comparison between outputs of dynamic system and simulated neural tree model and identification error: for training data set (upper) and for validation data set (lower)

comparison to tree-based GP lies in fact that computer programs are evolved at the machine code level, using lower level representations for the individuals. This can tremendously hasten up the evolution process as, no matter how an individual is initially represented, finally it always has to be represented as a piece of machine code, as fitness evaluation requires physical execution of the individuals. LGP based IDS is presented in [83].

Since the amount of audit data that an IDS needs to examine is very large even for a small network, analysis is difficult even with computer assistance because extraneous features can make it harder to detect suspicious behavior patterns. IDS must therefore reduce the amount of data to be processed. This is very important if real-time detection is desired. Reduction can occur in one of several ways. Data that is not considered useful can be filtered, leaving only the potentially interesting data. Data can be grouped or clustered to reveal hidden patterns; by storing the characteristics of the clusters instead of the data, overhead can be reduced. Finally, some data sources can be eliminated using feature selection. In the literature there are some related works for feature reduction in IDS. Support vector machine technique is used to select the important features [86]. Feature deduction using Markov blanket model and decision trees are presented in [76].

We illustrate a Flexible Neural Tree (FNT) [189] approach for selecting the input variables and detection of network intrusions. Based on the predefined instruction/operator sets, a flexible neural tree model can be created and evolved. FNT allows input variables selection, over-layer connections and different activation functions for different nodes. The hierarchical structure is evolved using tree-structure based evolutionary algorithm. The fine tuning of the parameters encoded in the structure is accomplished using particle swarm optimization (PSO) [136]. The proposed method interleaves both optimizations. Starting with random structures and corresponding parameters, it first tries to improve the structure and then as soon as an improved structure is found, it fine tunes its parameters. It then goes back to improving the structure again and, fine tunes the structure and rules' parameters. This loop continues until a satisfactory solution is found or a time limit is reached.

The Data Set

The data for the experiments was prepared by the 1998 DARPA intrusion detection evaluation program by MIT Lincoln Lab. The data set contains 24 attack types that could be classified into four main categories namely *Denial of Service (DOS)*, *Remote to User (R2L)*, *User to Root (U2R)* and *Probing*. The original data contains 744 MB data with 4,940,000 records. The data set has 41 attributes for each connection connection record plus one class label. Some features are derived features, which are useful in distinguishing normal from attacks. These features are either nominal or numeric. Some features examine only the connection in the past two seconds that have the same destination host as the current connection, and calculate statistics related

Table 2.4 Network Data Feature Labels

Label	Feature	Label	Feature
x_1	duration	x_2	protocol-type
x_3	service	x_4	flag
x_5	src_bytes	x_6	dst_bytes
x_7	land	x_8	wrong_fragment
x_9	urgent	x_{10}	hot
x_{11}	num_failed_logins	x_{12}	logged_in
x_{13}	num_compromised	x_{14}	root_shell
x_{15}	su_atempted	x_{16}	num_root
x_{17}	num_file_creations	x_{18}	num_shells
x_{19}	num_acess_files	x_{20}	num_outbound_cmds
x_{21}	is_host_login	x_{22}	is_guest_login
x_{23}	count	x_{24}	srv_count
x_{25}	serror_rate	x_{26}	srv_serror_rate
x_{27}	rerror_rate	x_{28}	srv_rerror_rate
x_{29}	smae_srv_rate	x_{30}	diff_srv_rate
x_{31}	srv_diff_host_rate	x_{32}	dst_host_count
x_{33}	dst_host_srv_count	x_{34}	dst_host_same_srv_rate
x_{35}	dst_host_diff_srv_rate	x_{36}	dst_host_same_srv_port_rate
x_{37}	dst_host_srv_diff_host_rate	x_{38}	dst_host_serror_rate
x_{39}	dst_host_srv_serror_rate	x_{40}	dst_host_rerror_rate
x_{41}	dst_host_srv_rerror_rate		

to protocol behavior, service, etc. These are called same host features. Some features examine only the connections in the past two seconds that have same service as the current connection and called same service features. Some other connection records were also stored by destination host, and features were constructed using a window of 100 connections to the same host instead of a time window. These are called host-based traffic features. R2L and U2R attacks don't have any sequential patterns like DOS and Probe because the former attacks have the attacks embedded in the data packets whereas the later attacks have many connections in a short amount of time. So some features that look for suspicious behavior in the data packets like number of failed logins are constructed and these are called contents features. The data used in the experiments reported in this Chapter contains randomly generated 11982 records having 41 features [87]. The labels of the 41 features and their corresponding networks data features are shown in Table 2.4.

This data set has five different classes namely *Normal DOS, R2L, U2R* and *Probe.* The training and test comprises of 5092 and 6890 records respectively. All the IDS models were trained and tested with the same set of data. As the data set has five different classes we performed a 5-class binary classification. The *normal* data belongs to Class 1, *Probe* belongs to Class 2, *DOS* belongs

Table 2.5 Parameters used in the flexible neural tree model

Parameter	Initial values
Population Size PS	100
Crossover probability	0.3
Opponent q in tournament selection	30
Maximum local search steps	2000
Terminate steps in local search	100
Initial connection weights	rand[-1, 1]
Initial parameters a_i and b_i	rand[0,1]

to Class 3, $U2R$ belongs to Class 4 and $R2L$ belongs to Class 5. The initial parameter settings used for each experiment are listed in Table 2.5.

Modeling IDS Using Flexible Neural Tree Using 41 Input-Variables

We used all the original 41 input variables for constructing a FNT model. A FNT classifier was constructed using the training data and then the classifier was used on the test data set to classify the data as an attack or normal data. The instruction sets used to create an optimal FNT classifier is $S = F \bigcup T = \{+_5, \ldots, +_{20}\} \bigcup \{x_1, x_2, \ldots, x_{41}\}$, where $x_i (i = 1, 2, \ldots, 41)$ denotes the 41 features.

The required number of iterations for structure and parameter optimization for each of the flexible neural tree classifiers are listed in Table 2.6.

Table 2.6 Iterations for structure and parameter optimization

Class	Structure optimization	Parameter optimization
Class 1	95	1789
Class 2	89	1602
Class 3	91	1539
Class 4	48	1892
Class 5	64	1920

The optimal flexible neural tree's for classes 1-5 are illustrated in Figures 2.9, 2.10 and 2.11. It should be noted that the important features for constructing the flexible neural tree model were formulated in accordance with the procedure mentioned in the previous Section. These important

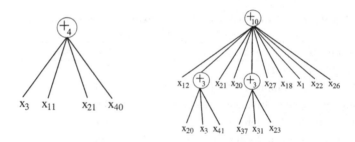

Fig. 2.9 The evolved FNT for Classes 1 and 2 with 41 input variables

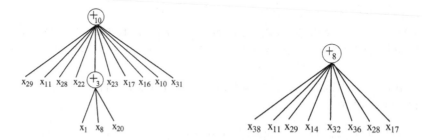

Fig. 2.10 The evolved FNT for Class 3 and Class 4 with 41 input variables

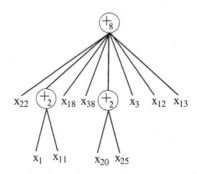

Fig. 2.11 The evolved FNT for Class 5 with 41 input variables

variables are shown in Table 2.7. Table 2.10 depicts the detection performance of the flexible neural tree by using the original 41 variable data set.

Modeling IDS Using Input Variables Selected by Decision Tree Approach

The important variables for intrusion detection were decided by their contribution to the construction of the decision tree [21–23, 76]. Variables

Table 2.7 The important features selected by the flexible neural tree algorithm

Class	Important variables
Class 1	x_3, x_{11}, x_{21}, x_{40}
Class 2	x_1, x_3, x_{12}, x_{18}, x_{20}, x_{21}, x_{23}, x_{26}, x_{27}, x_{31}, x_{37}, x_{41}
Class 3	x_1, x_8, x_{10}, x_{11}, x_{16}, x_{17}, x_{20}, x_{12}, x_{23}, x_{28}, x_{29}, x_{31}
Class 4	x_{11}, x_{14}, x_{17}, x_{28}, x_{29}, x_{32}, x_{36}, x_{38}
Class 5	x_1, x_3, x_{11}, x_{12}, x_{13}, x_{18}, x_{20}, x_{22}, x_{25}, x_{38}

rankings were generated in terms of percentages. The variables that had 0.00% rankings and considered only the primary splitters were eliminated [76]. This resulted in a reduced 12 variable data set with x_3, x_5, x_6, x_{12}, x_{23}, x_{24}, x_{25}, x_{28}, x_{31}, x_{32}, x_{33} and x_{35} as variables. Further the FNT classifier was constructed using the 12 variable data set (training data) and then the test data was passed through the save trained model. The instruction sets used to create an optimal neural tree model is $S = F \bigcup T = \{+_2, \ldots, +_{10}\} \bigcup \{x_3,$ x_5, x_6, x_{12}, x_{23}, x_{24}, x_{25}, x_{28}, x_{31}, x_{32}, x_{33}, $x_{35}\}$.

The iterations for structure and parameter optimization for each of the flexible neural tree classifiers are listed in Table 2.8.

Table 2.8 Number of iterations for structure and parameter optimization

Class	Structure optimization	Parameter optimization
Class 1	45	1745
Class 2	49	1403
Class 3	31	1547
Class 4	28	1678
Class 5	54	1834

The optimal flexible neural tree's for classes 1-5 are depicted in Figures 2.12, 2.13 and 2.14. It should be noted that the important features for constructing the flexible neural tree model were recognized automatically one more time. The important variables selected by the flexible neural tree model are illustrated in Table 2.9. Table 2.10 depicts the performance of the flexible neural tree by using the reduced 12 variable data set.

Modeling IDS Using Neural Networks without Input Variable Selection

For comparison purpose, a neural network classifier trained by PSO algorithm with flexible bipolar sigmoid activation functions were constructed using the same training data sets and then the neural network classifier was used on the

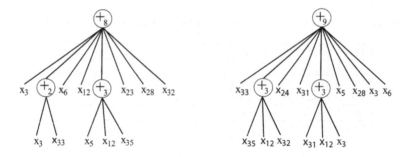

Fig. 2.12 The evolved FNT for Classes 1 and 2 with 12 input variables

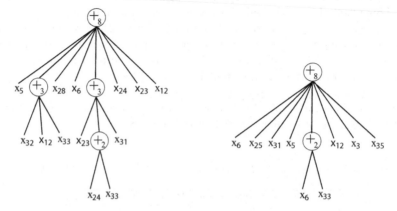

Fig. 2.13 The evolved FNT for Class 3 and Class 4 with 12 input variables

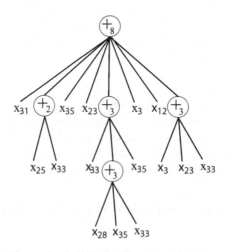

Fig. 2.14 The evolved FNT for Class 5 with 12 input variables

Table 2.9 Important features for 12 variable data set

Class	Important variables
Class 1	x_3, x_5, x_6, x_{12}, x_{23}, x_{28}, x_{32}, x_{33}, x_{35}
Class 2	x_3, x_5, x_6, x_{12}, x_{24}, x_{28}, x_{31}, x_{32}, x_{33}, x_{35}
Class 3	x_5, x_6, x_{12}, x_{23}, x_{24}, x_{28}, x_{31}, x_{32}, x_{33}
Class 4	x_3, x_5, x_6, x_{12}, x_{25}, x_{31}, x_{33}, x_{35}
Class 5	x_3, x_{12}, x_{23}, x_{25}, x_{28}, x_{31}, x_{33}, x_{35}

Table 2.10 Detection performance using flexible neural tree model using 41 and 12 input variables

Attack Class	41 variable data set	12 variable data set
Normal	**99.19%**	97.98%
Probe	**98.39%**	97.46%
DOS	**98.75%**	94.63%
U2R	99.70%	**99.76%**
R2L	**99.09%**	98.99%

test data set to detect the different types of attacks. All the input variables were used for the experiments.

Before describing details of the algorithm for training neural network classifier, the issue of coding is presented. Coding concerns the way the weights and the flexible activation function parameters of neural network are represented by individuals or particles. A float point coding scheme is adopted here. For neural network coding, suppose there are M nodes in hidden layer and one node in output layer and n input variables, then the number of total weights is $n * M + M * 1$, the number of thresholds is $M + 1$ and the number of flexible activation function parameters is $M + 1$, therefore the total number of free parameters in a neural network to be coded is $n * M + M + 2(M + 1)$. These parameters are coded into an individual or particle orderly.

The simple loop of the proposed training algorithm for neural network is as follows:

S1   Initialization. Initial population is generated randomly. The learning parameters c_1 and c_2 in PSO should be assigned in advance.
S2   Evaluation. The objective function value is calculated for each particle.
S3   Modification of search point.
S4   if maximum number of generations is reached or no better parameter vector is found for a significantly long time (100 steps), then stop, otherwise goto step **S2**;

Table 2.11 depicts the performance of the neural network by using original 41 variable data set and the 12 variable reduced data set.

Table 2.11 Detection performance using neural network classifier with 41 and 12 input variables

Attack Class	41 variable data set (original data)	12 variable data set (decision tree reduced data)
Normal	**95.69%**	95.59%
Probe	**95.53%**	95.08%
DOS	90.41%	**100%**
U2R	100%	100%
R2L	98.10%	**99.25%**

Table 2.12 The false positive/negative errors using 41 variable data set by the FNT algorithm

Attack Class	False positive error	False negative error
Normal	0.0581%	0.7837%
Probe	1.3943%	0.2160%
DOS	0.6241%	0.6241%
U2R	0.2177%	0.0726%
R2L	0.7547%	0.1597%

The false positive/negative errors using 41 variable data set by the flexible neural tree algorithm are illustrated in Table 2.12.

Time-Series Forecasting

Time-series forecasting is an important research and application area. Much effort has been devoted over the past several decades to develop and improve the time series forecasting models. Well established time series models include: (1)linear models, e.g., moving average, exponential smoothing and the autoregressive integrated moving average (ARIMA); (2)nonlinear models, e.g., neural network models and fuzzy system models. Recently a tendency for combining of linear and nonlinear models for forecasting time series has been an active research area [67].

The developed flexible neural tree model is illustrated in conjunction with two time-series prediction problems: Box-Jenkins time series and chaotic Mackey-Glass time-series. Well-known benchmark examples are used for the sake of an easy comparison with existing models. The data related to the examples are available on the web site of the Working Group on Data Modelling Benchmark-IEEE Neural Network Council [88].

For each benchmark problem, two experimental simulations are carried out. The first one use the same inputs with other models so as to make a meaningful comparison. The second one use a large number of input variables in order flexible neural tree to select proper input variables or time-lags automatically. In addition, the parameters used for each experiment is listed in Table 2.13.

Table 2.13 Parameters used in the flexible neural tree model

Parameter	Initial value
Population Size PS	30
Elitist Learning Probability P_{el}	0.01
Learning rate lr	0.01
Fitness constant ε	0.000001
Overall mutation probability P_M	0.4
Mutation rate mr	0.4
Prune threshold T_P	0.999999
Maximum local search steps	2000
Initial connection weights	rand[-1, 1]
Initial parameters a_i and b_i	rand[0,1]

Jenkins-Box Time-Series

The gas furnace data (series J) of Box and Jenkins (1970) was recorded from a combustion process of a methane-air mixture. It is well known and frequently used as a benchmark example for testing identification and prediction algorithms. The data set consists of 296 pairs of input-output measurements. The input $u(t)$ is the gas flow into the furnace and the output $y(t)$ is the CO_2 concentration in outlet gas. The sampling interval is 9s.

Case 1

The inputs for constructing flexible neural tree model are $u(t-4)$ and $y(t-1)$, and the output is $y(t)$. 200 data samples are used for training and the remaining data samples are used for testing the performance of evolved model.

The used instruction set for creating a flexible neural tree model is $S = F \bigcup T = \{+_2, +_3, \ldots, +_8\} \bigcup \{x_1, x_2\}$. Where x_1 and x_2 denotes the input variables $u(t-4)$ and $y(t-1)$, respectively.

After 37 generations of evolution, the optimal neural tree model was obtained with the MSE 0.000664. The MSE value for validation data set is 0.000701. The evolved neural tree is depicted in Figure 2.15 (left) and the actual time-series, the flexible neural tree model output and the prediction error is shown in Figure 2.15 (right).

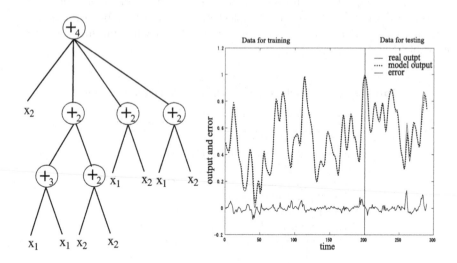

Fig. 2.15 Case 1: The evolved flexible neural tree model for prediction of Jenkins-Box data (left), and the actual time-series data, output of the evolved flexible neural tree model and the prediction error for training and test samples (right)

Case 2

For the second simulation, 10 inputs variables are used for constructing a FNT model. The proper time-lags for constructing a FNT model are finally determined by an evolutionary procedure.

The used instruction sets to create an optimal neural tree model is $S = F \bigcup T = \{+_2, \ldots, +_8\} \bigcup \{x_1, x_2, \ldots, x_{10}\}$. Where $x_i (i = 1, 2, \ldots, 10)$ denotes $u(t-6)$, $u(t-5)$, $u(t-4)$, $u(t-3)$, $u(t-2)$, $u(t-1)$ and $y(t-1)$, $y(t-2)$, $y(t-3)$ and $y(t-4)$ respectively.

After 17 generations of the evolution, the optimal neural tree model was obtained with the MSE 0.000291. The MSE value for validation data set is 0.000305. The evolved flexible neural tree is depicted in Figure 2.16 (left) and the actual time-series, the flexible neural tree model output and the prediction error is shown in Figure 2.16 (right). From the evolved flexible neural tree, it can be seen that the optimal inputs variables for constructing a flexible neural tree model are: $u(t-6)$, $u(t-5)$, $u(t-3)$, $y(t-1)$, $y(t-2)$, $y(t-3)$ and $y(t-4)$. It should be noted that the flexible neural tree model with proper selected input variables has accurate precision and good generalization ability. A comparison result of different methods for forecasting Jenkins-Box data is illustrated in Table 2.14.

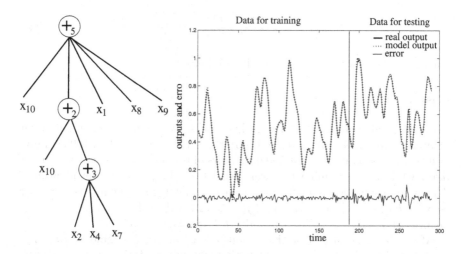

Fig. 2.16 Case 2: The evolved neural tree model for prediction of Jenkins-Box data (left), and the actual time series data, output of the evolved neural tree model and the prediction error for training and test samples (right)

Table 2.14 Comparative results of different modelling approaches

Model name and reference	Number of inuts	MSE
ARMA [89]	5	0.71
Tong's model [90]	2	0.469
Pedrycz's model [91]	2	0.320
Xu's model [92]	2	0.328
Sugeno's model [93]	2	0.355
Surmann's model [94]	2	0.160
TS model [95]	6	0.068
Lee's model [96]	2	0.407
Hauptmann's model [97]	2	0.134
Lin's model [98]	5	0.261
Nie's model [99]	4	0.169
ANFIS model [100]	2	0.0073
FuNN model [101]	2	0.0051
HyFIS model [102]	2	0.0042
FNT model (Case 1)	2	0.00066
FNT model (Case 2)	7	0.00029

Mackey-Glass Time-Series

The chaotic Mackey-Glass differential delay equation is recognized as a benchmark problem that has been used and reported by a number of researchers for comparing the learning and generalization ability of different models. The mackey-Glass chaotic time series is generated from the following equation:

$$\frac{dx(t)}{dt} = \frac{ax(t-\tau)}{1+x^{10}(t-\tau)} - bx(t). \tag{2.9}$$

where $\tau > 17$, the equation shows chaotic behavior.

Case 1

To make a fair comparison with earlier works, we predict the $x(t+6)$ with using the inputs variables $x(t)$, $x(t-6)$, $x(t-12)$ and $x(t-18)$. 1000 sample points were used and the first 500 data pairs of the series were used as training data, while the remaining 500 were used to validate the model identified. The used instruction sets to create an optimal flexible neural tree model is $S = F \bigcup T = \{+_5, \ldots, +_{10}\} \bigcup \{x_1, x_2, x_3, x_4\}$. Where $x_i(i = 1, 2, 3, 4)$ denotes $x(t)$, $x(t-6)$, $x(t-12)$ and $x(t-18)$, respectively.

After 135 generations of the evolution, an optimal flexible neural tree model was obtained with the RMSE 0.006901. The RMSE value for validation data set is 0.007123. The evolved flexible neural tree is depicted in Figure 2.17 (left). The actual time-series data, the output of flexible neural tree model and the prediction error are illustrated in Figure 2.17 (right). A comparison result of different methods for forecasting Mackey-Glass data is shown in Table 2.15.

Case 2

For the second simulation, 19 inputs variables are used for constructing a flexible neural tree model. The proper time-lags for constructing a flexible neural tree model are finally determined by an evolutionary procedure.

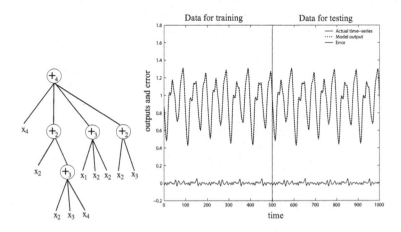

Fig. 2.17 Case 1: The evolved neural tree model for prediction of the Mackey-Glass time-series(left), and the actual time series data, output of the evolved neural tree model and the prediction error(right)

Table 2.15 Comparison results of the prediction error of different methods for the Mackey-Glass time-series problem

Method	Prediction error(RMSE)
Auto-regressive model	0.19
Cascade correlation NN	0.06
Back-propagation NN	0.02
Sixth-order polynomial	0.04
Linear prediction method	0.55
ANFIS and Fuzzy System [100]	0.007
Wang et al. Product T-norm [104]	0.0907
Classical RBF (with 23 neurons) [103]	0.0114
PG-RBF network [105]	0.0028
Genetic algorithm and fuzzy system [106]	0.049
FNT model (Case 1)	0.0069
FNT model (Case 2)	0.0027

The used instruction sets to create an optimal neural tree model is $S = F \bigcup T = \{+_2, \ldots, +_8\} \bigcup \{x_1, x_2, \ldots, x_{19}\}$. Where $x_i(i = 1, 2, \ldots, 19)$ denotes $x(t - 18)$, $x(t - 17)$, ..., $x(t - 1)$ and $x(t)$, respectively.

The optimal neural tree model was obtained with the RMSE 0.00271. The RMSE value for validation data set is 0.00276. The evolved flexible neural tree is shown in Figure 2.18 (left) and the actual time-series, the flexible neural tree model output and the prediction error is shown in Figure 2.18 (right). From the evolved flexible neural tree, it is evident that the optimal inputs

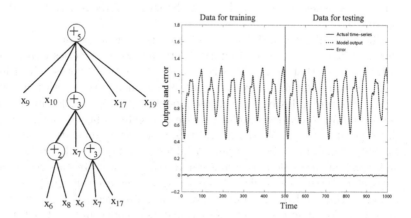

Fig. 2.18 Case 2: The evolved neural tree model for prediction of the Mackey-Glass time-series(left), and the actual time series data, output of the evolved neural tree model and the prediction error(right)

variables for constructing a flexible neural tree model are: $x(t-13)$, $x(t-12)$, $x(t-11)$, $x(t-10)$, $x(t-9)$, $x(t-2)$ and $x(t)$. That is, for predicting $x(t+6)$, among the time-lags from 0 to 18, the automatically evolved time-lags are 13, 12, 11, 10, 9, 2 and 0. It should be noted that the flexible neural tree model with proper selected time-lags as input variables has better precision and good generalization ability. A comparison result of different methods for forecasting Mackey-Glass data is illustrated in Table 2.15.

From above simulation results, it is evident that the proposed flexible neural tree model works well for generating prediction models of time series.

2.2.6   Exchange Rate Forecasting

Forecasting exchange rates is an important financial problem that is receiving increasing attention especially because of its difficulty and practical applications. Exchange rates are affected by many highly correlated economic, political and even psychological factors. These factors interact in a very complex fashion. Exchange rate series exhibit high volatility, complexity and noise that result from an elusive market mechanism generating daily observations [107][113][115][173].

Much research effort has been devoted to exploring the nonlinearity of exchange rate data and to developing specific nonlinear models to improve exchange rate forecasting, i.e., the autoregressive random variance (ARV) model [108], autoregressive conditional heteroscedasticity (ARCH) [109], self-exciting threshold autoregressive models [110]. There has been growing interest in the adoption of neural networks, fuzzy inference systems and statistical approaches for exchange rate forecasting problem [111][112][120][114][121] [122]. For a recent review of neural networks based exchange rate forecasting, the reader may consult [115].

The input dimension (i.e. the number of delayed values for prediction) and the time delay (i.e. the time interval between two time series data) are two critical factors that affect the performance of neural networks. The selection of dimension and time delay has great significance in time series prediction.

We illustrate a flexible neural tree model [189][190] for selecting the input variables and forecasting exchange rates. Based on the pre-defined instruction/operator sets, a flexible neural tree model can be created and evolved. Flexible neural tree allows input variables selection, over-layer connections and different activation functions for different nodes. In this research, the hierarchical structure of flexible neural tree is evolved using the Extended Compact Genetic Programming (ECGP), a tree-structure based evolutionary algorithm. The fine tuning of the parameters encoded in the structure is accomplished using particle swarm optimization (PSO). The proposed method interleaves both optimizations. Starting with random structures and corresponding parameters, it first tries to improve the structure and then as soon as an improved structure is found, it fine tunes its parameters. It then goes back to improving the structure again and, fine tunes the structure and rules' parameters. This loop continues until a satisfactory solution is found or a time limit is reached. The novelty of this paper is in the usage of flexible neural tree model for selecting the important inputs and/or time delays and for forecasting foreign exchange rates.

The Data Set

We used three different datasets in our forecast performance analysis. The data used are daily forex exchange rates obtained from the Pacific Exchange Rate Service [119], provided by Professor Werner Antweiler, University of British Columbia, Vancouver, Canada. The data comprises of the US dollar exchange rate against Euros, Great Britain Pound (GBP) and Japanese Yen (JPY). We used the daily data from 1 January 2000 to 31 October 2002 as training data set, and the data from 1 November 2002 to 31 December 2002 as evaluation test set or out-of-sample datasets (partial data sets excluding holidays), which are used to evaluate the good or bad performance of the predictions, based on evaluation measurements.

The forecasting evaluation criteria used is the normalized mean squared error (NMSE):

$$NMSE = \frac{\sum_{t=1}^{N}(y_t - \hat{y}_t)^2}{\sum_{t=1}^{N}(y_t - \bar{y}_t)^2} = \frac{1}{\sigma^2}\frac{1}{N}\sum_{t=1}^{N}(y_t - \hat{y}_t)^2, \qquad (2.10)$$

where y_t and $\hat{y}_t$ are the actual and predicted values, σ^2 is the estimated variance of the data and $\bar{y}_t$ the mean. The ability to forecast movement direction or turning points can be measured by a statistic developed by Yao and Tan [117]. Directional change statistics (Dstat) can be expressed as:

$$D_{stat} = \frac{1}{N}\sum_{t=1}^{N}a_t \times 100\%, \qquad (2.11)$$

where $a_t = 1$ if $(y_{t+1} - y_t)(\hat{y}_{t+1} - \hat{y}_t) \geq 0$, and $a_t = 0$ otherwise.

Feature/Input Selection Using Flexible Neural Tree

It is often a difficult task to select important variables for a forecasting or classification problem, especially when the feature space is large. A fully connected neural network classifier usually cannot do this. In the perspective of flexible neural tree framework, the nature of model construction procedure allows the flexible neural tree to identify important input features in building a forecasting model that is computationally efficient and effective. The mechanisms of input selection in the flexible neural tree constructing procedure are as follows. (1) Initially the input variables are selected to formulate the flexible neural tree model with same probabilities; (2) The variables which have more contribution to the objective function will be enhanced and have high opportunity to survive in the next generation by a evolutionary procedure; (3) The evolutionary operators i.e., crossover and mutation, provide a input selection method by which the flexible neural tree should select appropriate variables automatically.

Exchange Rate Forecasting - Experimental Results

For simulation, the five-day-ahead data sets are prepared for constructing flexible neural tree models. A flexible neural tree model was constructed using the training data and then the model was used on the test data set. The instruction sets used to create an optimal flexible neural tree forecaster is $S = F \bigcup T = \{+_2, +_3\} \bigcup \{x_1, x_2, x_3, x_4, x_5\}$, where $x_i (i = 1, 2, 3, 4, 5)$ denotes the 5 input variables of the forecasting model.

The optimal flexible neural tree's evolved for three major internationally traded currencies: British Pounds, Euros and Japanese Yen are illustrated in Figure 2.19. It should be noted that the important features for constructing the FNT model were formulated in accordance with the procedure mentioned in the previous Section.

For comparison purpose, the forecast performances of a traditional multilayer feed-forward network (MLFN) model and an adaptive smoothing neural network (ASNN) model are also illustrated in Tables 2.16 and 2.17. The actual daily exchange rates and the predicted ones for three major internationally traded currencies are depicted in Figures 2.20, 2.21 and 2.22. From Tables 2.16 and 2.17, it is observed that the proposed flexible neural tree forecasting models are better than other neural networks models for the three major internationally traded currencies.

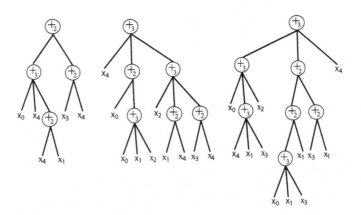

Fig. 2.19 The evolved FNT trees for forecasting euros (left), British pounds (middle) and Japanese yen (right)

Table 2.16 Forecast performance evaluation for the three exchange rates (NMSE for testing)

Exchange rate	euros	British pounds	Japanese yen
MLFN [120]	0.5534	0.2137	0.2737
ASNN [120]	0.1254	0.0896	0.1328
FNT (This paper)	0.0180	0.0142	0.0084

Table 2.17 Forecast performance evaluation for the three exchange rates (D_{stat} for testing)

Exchange rate	euros	British pounds	Japanese yen
MLFN [120]	57.5%	55.0%	52.5%
ASNN [120]	72.5%	77.5%	67.5%
FNT (This paper)	81.0	84.5%	74.5%

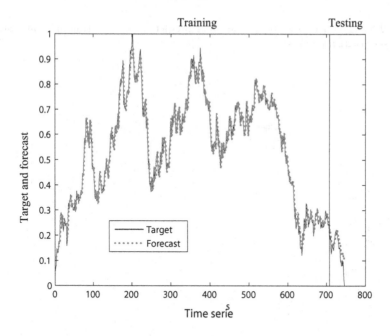

Fig. 2.20 The actual exchange rate and predicted ones for training and testing data set (euros)

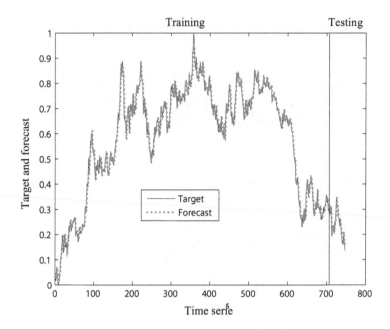

Fig. 2.21 The actual exchange rate and predicted ones for training dan testing data set (British pounds)

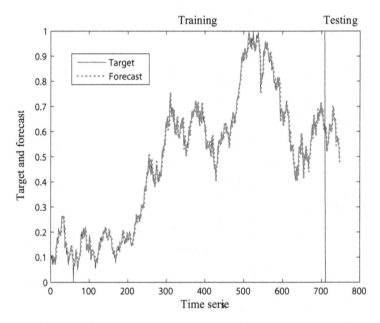

Fig. 2.22 The actual exchange rate and predicted ones for training dan testing data set (Japanese yen)

We have illustrated that the flexible neural tree forecasting model may provide better forecasts than the traditional MLFN and the ASNN forecasting models. The comparative evaluation is based on a variety of statistical measures such as NMSE and D_{stat}. Experimental analysis reveal that the NMSE and D_{stat} for three currencies using the FNT model are significantly better than those using the MLFN model and the ASNN model. This implies that the proposed FNT model can be used as a feasible solution for exchange rate forecasting.

2.2.7   Face Recognition

Face recognition has become a very active research area in recent years mainly due to increasing security demands and its potential commercial and law enforcement applications [156]. Face recognition is a very challenging problem and up to date, there is no technique that provides a robust solution to all situations and different applications that face recognition may encounter. Face recognition approaches on still images can be broadly grouped into geometric and template matching techniques. In the first case, geometric characteristics of faces to be matched, such as distances between different facial features, are compared. This technique provides limited results although it has been used extensively in the past. In the second case, face images represented as a two dimensional array of pixel intensity values are compared with a single or several templates representing the whole face. More successful template matching approaches use Principal Components Analysis (PCA) or Linear Discriminant Analysis (LDA) to perform dimensionality reduction achieving good performance at a reasonable computational complexity time. Other template matching methods use neural network classification and deformable templates, such as Elastic Graph Matching (EGM). Recently, a set of approaches that use different techniques to correct perspective distortion are being proposed. These techniques are sometimes referred to as view-tolerant. For a complete review on the topic of face recognition the reader may consult [123] and [124].

Neural networks have been widely applied in pattern recognition for the reason that neural-networks-based classifiers can incorporate both statistical and structural information and achieve better performance than the simple minimum distance classifiers [124]. Multilayered networks(MLNs), usually employing the backpropagation (BP) algorithm, are widely used in face recognition [129]. Recently, RBF neural networks have been applied in many engineering and scientific applications including face recognition [125][126].

We illustrate the use of a Flexible Neural Tree (FNT) [155][190][189] approach for selecting the input variables and face recognition. The hierarchical structure of the flexible neural tree is evolved using tree-structure based evolutionary algorithm. The fine tuning of the parameters encoded in the structure is accomplished using particle swarm optimization (PSO). The proposed method interleaves both optimizations. Starting with random structures and corresponding parameters, it first tries to improve the structure and then as soon as an improved structure is found, it fine tunes its parameters. It then goes back to improving the structure again and, fine tunes the structure and rules' parameters. This loop continues until a satisfactory solution is found or a time limit is reached.

Discrete Cosine Transform

Like other transforms, the Discrete Cosine Transform (DCT) attempts to decorrelate the image data [128]. After decorrelation each transform coefficient can be encoded independently without losing compression efficiency. The 2-D DCT is a direct extension of the 1-D case and is given by:

$$C(u,v) = \alpha(u)\alpha(v) \sum_{x=0}^{N-1} \sum_{y=0}^{N-1} f(x,y) cos\frac{\pi(2x+1)u}{2N} cos\frac{\pi(2y+1)}{2N} \quad (2.12)$$

for $u,v = 0,1,2,\ldots,N-1$ and $\alpha(u)$ and $\alpha(v)$ are defined as follows, $\alpha(u) = \sqrt{1/N}$ for $u = 0$, and $\alpha(u) = \sqrt{2/N}$ for $u \neq 0$.

The inverse transform is defined as:

$$f(x,y) = \sum_{u=0}^{N-1} \sum_{v=0}^{N-1} \alpha(u)\alpha(v) C(u,v) cos\frac{\pi(2x+1)u}{2N} cos\frac{\pi(2y+1)}{2N} \quad (2.13)$$

for $x,y = 0,1,2,\ldots,N-1$.

The DCT possess some fine properties, i.e., de-correlation, energy compaction, separability, symmetry and orthogonality. These attributes of the DCT have led to its widespread deployment in virtually every image/video processing standard of the last decade [128].

For an $M \times N$ image, we have $M \times N$ an DCT coefficient matrix covering all the spatial frequency components of the image. The DCT coefficients with large magnitude are mainly located in the upper-left corner of the DCT matrix. Accordingly, we scan the DCT coefficient matrix in a zig-zag manner starting from the upper-left corner and subsequently convert it to a one-dimensional (1-D) vector. Detailed discussions about image reconstruction errors using only a few significant DCT coefficients can be found in [127]. As a holistic feature extraction method, the DCT converts high-dimensional face images into low-dimensional spaces in which more significant facial features such as outline of hair and face, position of eyes, nose and mouth are maintained. These facial features are more stable than the variable high-frequency facial features. As a matter of fact, the human visual system is more sensitive to variations in the low-frequency band.

We also investigated the illumination invariant property of the DCT by discarding its several low-frequency coefficients. It is well-known that the first DCT coefficient represents the dc component of an image which is solely related to the brightness of the image. Therefore, it becomes DC free (i.e., zero mean) and invariant against uniform brightness change by simply removing the first DCT coefficient.

The ORL Face Database

The database consists of 400 images acquired from 40 persons with variations in facial expression (e.g. open / close eyes, smiling / non-smiling), and facial details (e.g. wearing glasses / not wearing glasses). All images were taken under a dark background, and the subjects were in an upright frontal position, with tilting and rotation tolerance up to 20 degree, and tolerance of up to about 10%. All images are of grey scale with a 92*112 pixels resolution. For each person, 5 images are generated randomly to form the training data set and the remaining were chosen as test data set.

Face Recognition Using Flexible Neural Tree with 55 Input-Variables

For this simulation, the DCT is employed to training and testing data sets, respectively. The extracted 55 input features are used for constructing a flexible neural tree model. A flexible neural tree classifier was constructed using the training data and then the classifier was used on the test data set to classify the data as an face ID or not. The instruction sets used to create an optimal flexible neural tree classifier is $S = F \bigcup T = \{+_5, \ldots, +_{20}\} \bigcup \{x_1, x_2, \ldots, x_{55}\}$, where $x_i (i = 1, 2, \ldots, 55)$ denotes the 55 features.

A comparison of different feature extraction methods and different face classification methods is illustrated in Table 2.18. Table 2.19 depicts the face recognition performance of the flexible neural tree by using the 55 features for test data set. The flexible neural tree method helps to reduce the features from 55 to 8-25.

Table 2.18 Comparison of different approaches for ORL face recognition

Method	Recognition rate
PCA+RBFN [130]	94.5%
LDA+RBFN [130]	94.0%
FS+RBFN [130]	92.0%
DCT+FNT (FNT)	98.13%

Table 2.19 The true positive (tp), false positive (fp) rates for flexible neural tree face recognition

Face ID	TP	FP	Accuracy
S1	80.0%	1.54%	98.0%
S2	80.0%	1.54%	98.0%
S3	80.0%	4.61%	95.0%
S4	100.0%	2.56%	97.5%
S5	80.0%	0.51%	99.0%
S6	100.0%	4.10%	96.0%
S7	40.0%	0.51%	98.0%
S8	80.0%	0.51%	99.0%
S9	80.0%	1.53%	98.0%
S10	100.0%	0.00%	100.0%
S11	60.0%	3.59%	95.5%
S12	80.0%	0.00%	99.5%
S13	80.0%	0.51%	99.0%
S14	80.0%	0.51%	99.0%
S15	80.0%	3.07%	96.5%
S16	80.0%	0.51%	99.0%
S17	100.0%	4.62%	95.5%
S18	100.0%	0.51%	99.5%
S19	100.0%	2.05%	98.0%
S20	60.0%	0.00%	99.0%
S21	100.0%	0.00%	100.0%
S22	80.0%	0.00%	99.5%
S23	80.0%	0.51%	99.0%
S24	80.0%	2.05%	97.5%
S25	100.0%	2.05%	98.0%
S26	80.0%	0.51%	99.0%
S27	60.0%	1.02%	98.0%
S28	60.0%	1.54%	97.5%
S29	100.0%	3.07%	97.0%
S30	60.0%	0.00%	99.0%
S31	100.0%	0.51%	99.5%
S32	80.0%	1.03%	98.5%
S33	100.0%	0.51%	99.5%
S34	100.0%	0.51%	99.5%
S35	60.0%	2.05%	97.0%
S36	60.0%	2.05%	97.0%
S37	80.0%	1.54%	98.0%
S38	80.0%	3.07%	96.5%
S39	100.0%	0.51%	99.5%
S40	80.0%	4.62%	95.0%
Average	82.0%	1.50%	98.13%

2.2.8   Microarray-Based Cancer Classification

The classification of cancers from gene expression profiles is actively investigated in bioinformatics. It commonly consists of feature selection and pattern classification. In advance, feature selection selects informative features useful to categorize a sample into predefined classes from lots of gene expression profiles. Pattern classification is composed of learning a classifier with those features and categorizing samples with the classifier.

Much research effort has been devoted to exploring the informative gene selection from microarray data. Typical effective feature reduction methods include principal components analysis (PCA), class-separability measure, Fisher ratio and t-test. Evolutionary based feature selection methods are alternatives of the gene selection approaches. A probabilistic model building genetic algorithm based informative selection method was proposed in [131]. Genetic programming can be also used to select informative gene and classification of gene expression profiles [132]. After the gene selection was performed, many candidate classifiers can be employed for classification of microarray data, including Bayessian network, KNN, neural networks, support vector machine [139], random forest [134] etc. For a recent review, the reader may consult [133]. Classification algorithms that directly provide measures of variable importance are of great interest for gene selection, specially if the classification algorithm itself presents features that make it well suited for the types of problems frequently faced with microarray data. Random forest is one such algorithm [134] and flexible neural tree method is another alternative.

We illustrate a flexible neural tree model [189][190] is employed for selecting the input variables and detecting the cancers. Based on the pre-defined instruction/operator sets, a flexible neural tree model can be created and evolved. Flexible neural tree allows input variables selection, over-layer connections and different activation functions for different nodes. The hierarchical structure is evolved using the Extended Compact Genetic Programming (ECGP), a tree-structure based evolutionary algorithm [191]. The fine tuning of the parameters encoded in the structure is accomplished using particle swarm optimization (PSO). The proposed method interleaves both optimizations. Starting with random structures and corresponding parameters, it first tries to improve the structure and then as soon as an improved structure is found, it fine tunes its parameters. It then goes back to improving the structure again and, fine tunes the structure and rules' parameters. This loop continues until a satisfactory solution is found or a time limit is reached. The novelty of this paper is in the usage of flexible neural tree model for selecting the informative genes and for classification of microarray data.

Feature/Input Selection Using Flexible Neural Tree

It is often a difficult task to select important variables for a forecasting or classification problem, especially when the feature space is large. A fully connected neural network classifier usually cannot do this. In the perspective of flexible neural tree framework, the nature of model construction procedure allows the flexible neural tree to identify important input features in building a forecasting model that is computationally efficient and effective. The mechanisms of input selection in the flexible neural tree construction procedure are as follows:

- Initially the input variables are selected to formulate the flexible neural tree model with same probabilities;
- The variables which have more contribution to the objective function will be enhanced and have high opportunity to survive in the next generation by a evolutionary procedure;
- The evolutionary operators i.e., crossover and mutation, provide a input selection method by which the flexible neural tree should select appropriate variables automatically.

Data Sets

The colon cancer dataset contains gene expression information extracted from DNA microarrays [135]. The dataset consists of 62 samples in which 22 are normal samples and 40 are cancer tissue samples, each having 2000 features. We randomly choose 31 samples for training set and the remaining 31 samples were used as testing set. (Available at: http://sdmc.lit.org.sg/GEDatasets/ Data/ColonTumor.zip). The leukemia dataset consists of 72 samples divided into two classes ALL and AML [140]. There are 47 ALL and 25 AML samples and each contains 7129 features. This dataset was divided into a training set with 38 samples (27 ALL and 11 AML) and a testing set with 34 samples (20 ALL and 14 AML) (Available at: http://sdmc.lit.org.sgGEDatasets DataALL-AML_Leukemia.zip).

Colon Cancer

The data was randomly divided into a training set of 30 samples and testing set of 12 for 50 times, and the final results were averaged over these 30 independent trials. A model was constructed using the training data and then the model was used on the test data set. The instruction sets used to create an optimal flexible neural tree model is $S = F \bigcup T = \{+_5, +_6, \ldots, +_9\} \bigcup \{x_0, x_1, \ldots, x_{1999}\}$, where $x_i(i = 0, 1, \ldots, 1999)$ denotes the 2000 input variables (genes) of the classification model.

A best flexible neural tree tree obtained for Colon cancer is illustrated in Figure 2.23. It should be noted that the important features for constructing the flexible neural tree model were formulated in accordance with the procedure mentioned in the previous section. These informative genes selected by flexible neural tree algorithm is shown in Table 2.20.

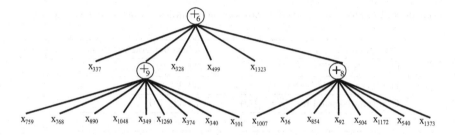

Fig. 2.23 An evolved best flexible neural tree for colon data classification

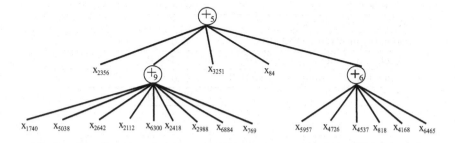

Fig. 2.24 An evolved best flexible neural tree for leukemia data classification

Table 2.20 The extracted informative genes in case of Colon dataset

$x_{337}, x_{328}, x_{759}, x_{768}, x_{890}, x_{1048}, x_{349}, x_{1260}, x_{374}, x_{340}, x_{101},$
$x_{499}, x_{1007}, x_{36}, x_{854}, x_{92}, x_{504}, x_{1172}, x_{540}, x_{1373}, x_{1323}$

Table 2.21 The extracted informative genes in case of leukemia dataset

$x_{2356}, x_{3251}, x_{1740}, x_{5038}, x_{2642}, x_{2112}, x_{6300}, x_{2418}, x_{2988}, x_{6884}, x_{769},$
$x_{5957}, x_{4726}, x_{4537}, x_{818}, x_{4168}, x_{6465}, x_{84}$

For comparison purpose, the classification performances of a genetic algorithm trained SVM [137], Bootstrapped GA+SVM [138], Combined kernel for SVM [139] and the flexible neural tree method are shown in Table 2.22. It is observed that the proposed flexible neural tree classification models are better than other models for classification of microarray dataset.

Leukemia Cancer

Leukemia dataset is divided into training and testing set. To setup the 30 independent trials, A flexible neural tree model was constructed using the

Table 2.22 The best prediction rate of some studies in case of Colon dataset

Classifier	Classification rate (%)
GA+SVM [137]	84.7± 9.1
Bootstrapped GA+SVM [138]	80.0
Combined kernel for SVM [139]	75.33±7.0
FNT	97.09±0.018

Table 2.23 The best prediction rate of some studies in case of Colon dataset

Classifier	Classification rate (%)
Weighted voting [135]	94.1
Bootstrapped GA+SVM [138]	97.0
Combined kernel for SVM [139]	85.3±3.0
Multi-domain gating network [197]	75.0
FNT	99.6±0.021

training data and then the model was used on the test data set. The instruction sets used to create an optimal flexible neural tree model is $S = F \bigcup T = \{+_5, +_6, \ldots, +_9\} \bigcup \{x_0, x_1, \ldots, x_{7128}\}$, where $x_i (i = 0, 1, \ldots, 7128)$ denotes the 7129 input variables (genes) of the classification model.

A best flexible neural tree tree obtained for leukemia cancer classification is depicted in Figure 2.24. It should be noted that the important features for constructing the flexible neural tree model were formulated in accordance with the procedure mentioned in the previous Section. These informative genes selected by flexible neural tree algorithm is illustrated in Table 2.21.

For comparison purposes, the classification performances of Weighted voting method [135], Bootstrapped GA+SVM [138], Combined kernel for support vector machines (SVM) [139], Multi-domain gating network [197] and the flexible neural tree method are shown in Table 2.23. It is observed that the proposed flexible neural tree classification models are better than other models for classification of microarray dataset.

2.2.9   Protein Fold Recognition

Protein structure classification represents an important process in understanding the associations between sequence and structure as well as possible functional and evolutionary relationships. Recent structural genomics initiatives and other high-throughput experiments have populated the biological databases at a rapid pace. The amount of structural data has made traditional methods such as manual inspection of the protein structure become impossible.

Protein fold recognition is an important approach to structure discovery without relying on sequence similarity. For the past several decades several methods have been proposed for predicting protein structural classes. These methods include discriminant analysis [141], correlation coefficient [142], hydrophobicity profiles [143], amino acid index [144], Bayes decisions rule [145], amino acid distributions [146], functional domain occurrences [147], supervised fuzzy clustering approach [154] and amino acid principal component analysis [149] etc.

On the other hand, alignment profiles are widely used for recognizing protein folds. Recently, Cheng and Baldi proposed a machine learning algorithm using secondary structure, solvent accessibility, contact map and strand pairing for fold recognition, which showed the pairwise sensitivity of 27% [150]. For the protein fold prediction problem, it has been reported that the amino acid properties are the key determinants of protein folding and are used for discriminating membrane proteins, identification of membrane spanning regions, prediction of protein structural classes, protein folding rates, protein stability[153] etc. Towards this direction, Ding and Dubchak [148] proposed a method based on neural networks and support vector machines for fold recognition using amino acid composition, and reported a cross-validated sensitivity of 45%. Taguchi and Gromiha [151] used the amino acid occurrence (not composition) of proteins belonging to 30 major folds and four structural classes based on linear discriminant analysis (LDA) and got an accuracy of 37% for recognizing 1612 proteins from 30 different folds.

We illustrate a flexible neural tree approach for multi-class protein fold classification problem. The approach includes four parts: feature selection, ECOC, SVM and compare system. The dataset we used for training was selected from the dataset built for the prediction of 128 folds in an earlier study [152]. This database was based on the PDB_select sets where two proteins have no more than 35% of the sequence identity for the aligned subsequences longer than 80 residues. We utilized 27 most populated folds in the database which have seven or more proteins and represent all major structural classes. As an independent dataset for testing we used the PDB_40D set developed by the authors of the SCOP database. This set contains the SCOP sequences having less than 40% identity with each other. From this set we selected 386 representatives of the same 27 largest folds. All PDB_40D proteins that had higher than 35% identity with the proteins of the training set were excluded from the testing set. We get the best prediction accuracy of 62.86% based on feature selection, ECOC and SVM, a little better than the accuracy of 56.5% while using support vector machine and neural networks.

Data Set

The data used in the experiment is the SCOP database having 27 classes proteins with 313 training samples and 385 testing samples. These samples have six features: amino acid composition, polarity, polarizability, normalized van der waals volume, hydrophobicity and predicted secondary structure (see Table 2.24).

Experimental Illustrations

The result of every SVM is illustrated in Table 2.25 and the final recognition rate is obtained by using the comparison system (see Table 2.26.). From

Table 2.24 The Description of fold feature

Symbol	Feature Name	Dimension
C	Amino Acid Composition	20
S	Predicted Secondary Structure	21
H	Hydrophobicity	21
P	Polarity	21
V	Normalized Van Der Waals Volume	21
Z	Polarizability	21

Table 2.25 The results for each support vector machine classifier

Index	Feature Name	λ	Training rate(%)	Testing rate (%)
0	CSHPV	1.0	100.0	74.8052
1	C	0.6	100.0	77.4026
2	CS	0.8	100.0	77.4026
3	C	0.6	100.0	80.2597
4	CSHPV	1.0	100.0	77.4026
5	CS	0.8	100.0	78.961
6	C	0.6	100.0	77.9221
7	CSHPV	1.0	100.0	74.2587
8	CSHPV	1.0	100.0	76.3636
9	C	0.6	100.0	76.8831
10	C	0.6	100.0	75.5844
11	CSHPV	1.0	100.0	78.7013
12	CSHPV	1.0	100.0	78.4416
13	CSHPV	1.0	100.0	81.039
14	C	1.0	100.0	79.2208
15	CSHPV	1.0	100.0	76.1039
16	CS	0.8	100.0	80.7792
17	C	0.6	100.0	79.4805
18	CSHPV	1.0	100.0	84.6753
19	CS	0.8	100.0	81.8182
20	CSHPV	1.0	100.0	78.4416
21	CSHPV	1.0	100.0	85.1948
22	C	0.6	100.0	78.1818
23	CSHPV	1.0	100.0	79.2208

Table 2.26 The comparison of independent test

Index	Rate(%)(FNT)	Ding's Rate [148](%)
1	83.3	83.3
3	77.8	77.8
4	60.0	35.0
7	37.5	50.0
9	100.0	100.0
11	55.6	66.7
20	84.1	71.6
23	33.3	16.7
26	76.9	50.0
30	33.3	33.3
31	37.5	50.0
32	31.6	26.3
33	75.0	50.0
35	25.0	25.0
39	85.7	57.1
46	93.8	77.1
47	58.3	58.3
48	61.5	48.7
51	40.7	61.1
54	33.3	36.1
57	62.5	50.0
59	50.0	35.7
62	57.1	71.4
69	25.0	25.0
72	37.5	12.5
87	29.6	37.0
110	96.3.	83.3
Q/%	62.86	56.0

simulation results, it is evident that the best prediction accuracy of 62.86% can be achieved by using the flexible neural tree method. This result is better than the methods of support vector machine and neural networks as reported in [148].

Multi-class protein folds recognition becomes more and more important in recent years, especially with the rapid development of bioinformation. We proposed an error-correcting output coding and SVM to recognize the multiclass protein fold. The experimental results shown that the best prediction accuracy of 62.86% can be achieved by the proposed method.

2.3 Multi Input Multi Output Flexible Neural Tree

In the past few years, much success has been achieved in the use of Flexible Neural Trees (FNT) for classification [194], recognition [234], approximation

[157][158] and control [159]. It has been shown that flexible neural tree is capable of approximating not only a continuous function but also its derivatives to an arbitrary degree of accuracy.

We illustrate a multi-input multi-output FNT (MIMO-FNT) model for the identification of nonlinear systems. Based on pre-defined instruction (operator) sets, the MIMO-FNT model can be created and evolved, in which over-layer connections, different activation functions for different nodes (neurons) are allowed. Therefore, the MIMO-FNT model can be viewed as a kind of irregular multi-layer flexible neural network. The MIMO-FNT structure is developed using the Immune Programming and the free parameters embedded in the neural tree are optimized by particle swarm optimization algorithm.

2.4   Representation and Calculation of the MIMO FNT

The commonly used representation (encoding) methods for an artificial neural networks are direct encoding scheme and indirect encoding scheme. The former uses a fixed structure (connection matrix or bit-strings) to specify the architecture of a corresponding neural network. The latter uses rewrite rules, i.e., cellular encoding and graph generation grammars to specify a set of construction rules that are recursively applied to yield an artificial neural network.

In a flexible neural tree representation, there is no additional effort needed to encode and decode between the genotype and the phenotype of a neural network. This is due to the fact that a neural tree can be directly calculated as a flexible neural network. Therefore, the direct neural tree representation reduces some computational expenses when calculating the object function. It suffers a possible problem in terms of search space, the search space may becomes larger if there is no additional technique to limit the architecture of neural tree. It is illustrated in [66], that based on a simple computation, the size of the sigma-pi neural tree structure space is $2^{31} * 5^{125}$, if 2 function operators $F = \{\Sigma, \Pi\}$, 5 terminals $T = \{x_1, x_2, x_3, x_4, x_5\}$, and maximum width 5 and maximum depth 3 are used. It is clear that any exhaustive search method for finding an optimal solution is impossible.

To cope with the huge search space problem and make the use of expert's knowledge, we illustrate a method in which the instructions of root node, hidden nodes and input nodes are selected from the following instruction set.

$$I = \{+_2, +_3, \ldots, +_N, x_0, x_1, \ldots, x_n\} \tag{2.14}$$

Based on the above instruction set, an example of the flexible neural tree is depicted in Figure 2.25, which contains input, hidden and output layers. In this case, the user can freely choose the instruction sets according to a given problem, i.e., selecting a bigger N_0 and a smaller N_1 should create a neural

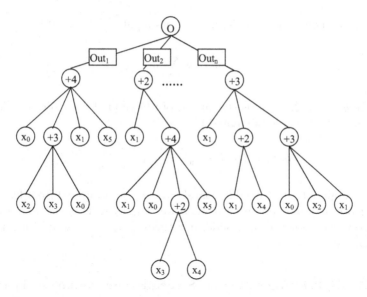

Fig. 2.25 A typical representation of neural tree for MIMO with three instruction sets and six input variables x_0, x_1, ..., x_5

tree with bigger width in the root node (bigger number of hidden neurons) and smaller width in hidden nodes.

Note that the instructions of output-layer (root node), hidden-layer, and input-layer are selected from instruction set I_0 with the specific probabilities, respectively.

Where nonterminal instruction $+_N$ has N arguments and the terminal instruction has no arguments. In the construction of a neural tree, if a nonterminal instruction, i.e., $+_p (p = 2, 3, 4, \ldots)$ is selected, p weights are created randomly and used for representing the connection strength between the node $+_p$ and its children. In addition, two parameters a_p and b_p are randomly created as flexible activation function parameters and attached to node $+_p$. A candidate flexible activation function used in the experiments is as follows:

$$f(a, b; x) = exp(-(\frac{x - a}{b})^2) \tag{2.15}$$

Let $I_{d,w}$ denote the instruction of a node with depth d and width w. The probabilities of selecting instructions for output layer, hidden layer and input layer in the instruction set I are initialized as follows:

$$P(I_{d,w}) = \frac{1}{l}, \forall I_{d,w} \in I \tag{2.16}$$

where l is the number of instructions in the instruction set I.

For any nonterminal node, i.e., $+_p$, its input is calculated by:

$$net_p = \sum_{i=1}^{p} w_i * y_i \tag{2.17}$$

where $y_i (i = 1, 2, \ldots, p)$ is the output of i-*th* child of the node $+_p$. Then, the output of node $+_p$ is calculated by:

$$out_p = f(a, b; net_p) = exp(-(\frac{net_p - a_p}{b_p})^2) \tag{2.18}$$

The overall output of flexible neural tree can be computed from left to right by depth-first method recursively. Note that there is no flexible activation function used for the root node, it returns a weighted sum of a number of nonlinear terms only.

2.4.1   Hybrid Algorithm for Structure and Parameter Learning

Structure Optimization

To optimize the structure of the flexible neural tree model, the Immune Programming (IP) is used. The algorithm of IP is based on the concept of evolving a repertoire of antibodies that encode candidate solutions to a given problem. At the start, candidate solutions to the program are randomly generated providing an initial repertoire of adequate diversity. The evolution of the repertoire is driven by cloning, mutation, and replacement of the antibodies. These processes maintain the diversity of the repertoire and expand the space searched for solutions. The algorithm is briefly described as follows.
1. Initialization. An initial repertoire (population), AB, of n antibodies, Ab_i, $(i = 1, \ldots, n)$ is generated. The generation counter is set to $G = 1$.
   2. Evaluation. An antigen, Ag, representing the problem to be solved, is presented. Ag is compared to all antibodies $Ab_i \in AB$ and their affinity, f_i, with respect to the antigen is determined.
   3. Replacement. With a certain probability, P_r, a new antibody is generated and placed into the new repertoire. This way, low affinity antibodies are implicitly replaced. The parameter P_r is the probability of replacement.
   4. Cloning. If a new antibody has not been generated, an antibody, Ab_i, is drawn form the current repertoire with a probability directly proportional to its antigenic affinity. With a probability, P_c, this antibody is cloned and placed in the new repertoire. The parameter P_c is termed probability of cloning.
   5. Mutation. If the high-affinity antibody selected in the previous step has not been cloned, it is submitted for mutation with a probability inversely

proportional to its antigenic affinity. If the antibody is selected for mutation, each component of its attribute string mutated with the probability of
mutation P_m.

6. Iteration-repertoire. Steps 3-5 are replaced until a new repertoire AB'
of size n is constructed.

7. Iteration-algorithm. The generation counter is incremented, $G = G + 1$,
and the new repertoire is submitted to step 2, evaluation. The process continues iteratively until a stopping criteria is met.

Parameter Optimization

For learning the parameters (weights and activation parameters) of a neural
tree model, there are a number of learning algorithms, such as genetic algorithm, evolutionary programming, gradient descent based learning method
and so on, that can be used for tuning of the parameters. The particle swarm
optimization (PSO) algorithm conducts search using a population of particles
that correspond to individuals in an Evolutionary Algorithm (EA). Initially,
a population of particles is randomly generated. Each particle represents a
potential solution and has a position represented by a position vector x_i.
Aswarm of particles moves through the problem space with the moving velocity of each particle represented by a velocity vector v_i. At each time step,
a function f_i - representing a quality measure - is calculated by using x_i as
input. Each particle keeps track of its own best position, which is associated
with the best fitness it has achieved so far in a vector p_i. Furthermore, the
best position among all the particles obtained so far in the population is kept
track of as p_g. In addition to this global version, another version of PSO keeps
track of the best position among all the topological neighbors of a particle. At
each time step t, by using the individual best position, $p_i(t)$, and the global
best position, $p_g(t)$, a new velocity for particle i is updated by.

$$V_i(t + 1) = v_i(t) + c_1\phi_1(p_i(t) - x_i(t))$$
$$+ c_2\phi_2(p_g(t) - x_i(t)) \tag{2.19}$$

where c_1 and c_2 are positive constants and ϕ_1 and ϕ_2 are uniformly distributed
random numbers in [0,1]. The term c_i is limited to the range of V_{max} (if the
velocity violates this limit, it is set to its proper limit). Changing velocity this
way enables the particle i to search around both its individual best position,
p_i, and global best position, p_g. Based on the updated velocities, each particle
changes its position according to:

$$x_i(t + 1) = x_i(t) + v_i(t + 1) \tag{2.20}$$

2.4.2 Hybrid Algorithm for Flexible Neural Tree Model

To find a good flexible neural tree model, structure and parameters optimization is both used in the hybrid algorithm. Combined the given IP and PSO algorithm, the hybrid algorithm for flexible neural tree model is described below.

Step 1. Parameters Definition. Before the algorithm, many parameters should be given first, i.e., size of population, size of agent, and so on.

Step 2. Initialization. Create N flexible neural tree model $A(t)$ randomly limited by the given parameters. Set $t = 0$.

Step 3. Weights Optimization by PSO Algorithm. For each flexible neural tree model $(A_0(t), A_1(t), ..., A_N(t))$, Optimize the weights by the PSO algorithm.

Step 4. Structure Optimization by IP. Create the new population $A(t + 1)$ by IP. Set $t = t + 1$.

Step 5. Iteration. The new repertoire is submitted to Step 2, the process continues iteratively until a stopping criteria is met.

2.4.3 Illustrative Examples

Single Input and Single Output (SISO) nonlinear system and Multi Input and Multi Output (MIMO) nonlinear system [160] are used. For each benchmark problem of the following examples, the instruction sets are selected according to the following ideas: the instruction set I is selected containing all terminal instructions $\{x_0, x_1, \ldots, x_n\}$ and additional nonterminal instructions $\{+_2, \ldots, +_p\}$, here p is a user defined integer number. Note that if p is large than the input number, according to our experiments, the node in the neural tree may contain redundant input information and create some difficulties for the training process, therefore we usually select the p within the interval $[1, numberofinputs]$.

Nonlinear System with Single Input Single Output

In this example, a nonlinear system with single input and single output is considered, the model has the following form:

$$y(k + 1) = \frac{y(k)}{1 + y^2(k)} + u^3(k) \tag{2.21}$$

where $u(k)$ is the external input for the system and $y(k)$ is the output of the system.

A training and a test sequence each were generated. The training sequence consists of 2000 samples in which u(k) is a random input in the interval [-2, 2]

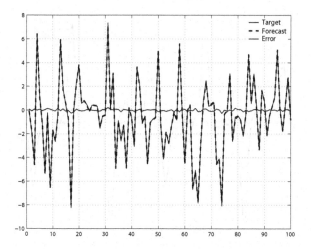

Fig. 2.26 Training sequence of the single input and single output nonlinear system

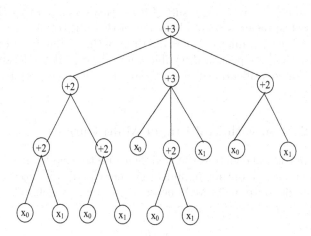

Fig. 2.27 Flexible neural tree model for the single input and single output non-linear system

(Please see Figure 2.26). The test sequence is made up of 100 samples by the external input $u(k) = sin(2\pi k/25) + sin(2\pi k/10)$. The used instruction set is $I = \{+_2, +_3, x_0, x_1\}$. where x_0, x_1 represents $u(k)$ and $y(k)$ respectively. The output is $y(k + 1)$.

Using the hybrid algorithm described in section 3, a flexible neural tree model model is found (Figure 2.27) and the identification result is illustrated in Figure 2.28.

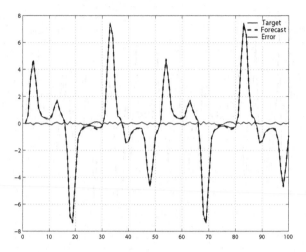

Fig. 2.28 Test sequence of the single input and single output nonlinear system

In this example, the training result of Root Mean Square Error (RMSE) is 0.088 while its test result is 0.067 a little better than the training one. The test forecast curve is very similar to the real output of the system from Figure 2.28 which showed that the proposed flexible neural tree model is feasibility and effectiveness for identification the single input and single output nonlinear dynamic system.

Nonlinear System with Multi Input Multi Output

In this example, it is shown that the flexible neural tree model used to identify single input and single output nonlinear system can also be used to identify multi input multi output (MIMO) nonlinear system as well. The system is described by the following equations:

$$y_1(k+1) = \frac{y_1(k)}{1 + y_2^2(k)} + u_1(k) \tag{2.22}$$

$$y_2(k+1) = \frac{y_1(k)y_2(k)}{1 + y_2^2(k)} + u_2(k) \tag{2.23}$$

where $u_1(k)$ and $u_2(k)$ is the external input for the system, $y_1(k)$ and $y_1(k)$ is the output of the system.

A training and a test sequence each were generated. The training sequence consists of 5000 samples in which $u_1(k)$ and $u_2(k)$ are two random inputs in

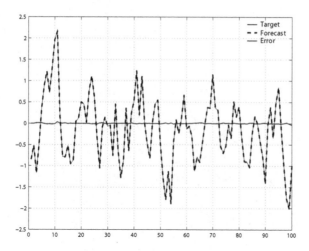

Fig. 2.29 Training sequence of y_1 of the multi input multi output nonlinear system

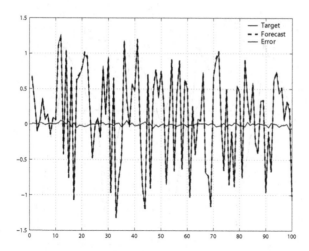

Fig. 2.30 Training sequence of y_2 of the multi input multi output nonlinear system

the interval [-1, 1] (Please see Figures 2.29 and 2.30). The test sequence is made up of 100 samples by the external inputs $u_1(k) = sin(2\pi k/25)$ and $u_2(k) = cos(2\pi k/25)$. The used instruction sets are $I = \{+_2, +_3, \ldots, +_5, x_0, x_1, x_2, x_3\}$, where x_0, x_1 represents $y_1(k)$, $y_2(k)$ and x_3, x_4 represents $u_1(k)$, $u_2(k)$ respectively. The two outputs of the system are $y_1(k+1)$ and $y_2(k+1)$.

In this example, the training results of Root Mean Square Error (RMSE) are 0.0091 (y_1) and 0.0224 (y_2) while the test results are 0.2205 (y_1) and

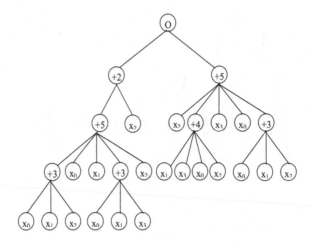

Fig. 2.31 Flexible neural tree model for the multi input multi output nonlinear system

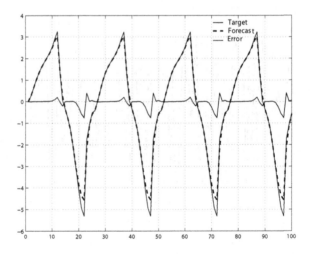

Fig. 2.32 Test sequence of y_1 of the multi input multi output nonlinear system

0.2580 (y_2). The optimized multi input multi output-flexible neural tree model is shown in Figure 2.31 and the identification results are shown in Figures 2.32 and 2.33, respectively.

From above simulation results, it can be seen that the proposed neural tree model works very well for the nonlinear dynamic system identification problems.

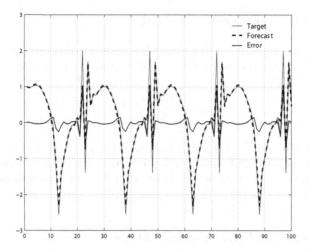

Fig. 2.33 Test sequence of y_2 of the multi input multi output nonlinear system

2.5   Ensemble of Flexible Neural Tree

For most regression and classification problems, combining the outputs of several predictors improves on the performance of a single generic one [182]. Formal support to this property is provided by the so-called bias/variance dilemma [169], based on a suitable decomposition of the prediction error. According to these ideas, good ensemble members must be both accurate and diverse, which poses the problem of generating a set of predictors with reasonably good individual performances and independently distributed predictions for the test points. Diverse individual predictors can be obtained in several ways. These include:

(i)   using different algorithms to learn from the data (classification and regression trees, artificial neural networks, support vector machines, etc.);
(ii) changing the internal structure of a given algorithm (for instance, number of nodes/depth in trees or architecture in neural networks);
(iii) learning from different adequately- chosen subsets of the data set.

The probability of success in strategy (iii), the most frequently used, is directly tied to the instability of the learning algorithm [163]. That is, the method must be very sensitive to small changes in the structure of the data and/or in the parameters defining the learning process. Again, classical examples in this sense are classification and regression trees and artificial neural networks. In particular, in the case of artificial neural networks the instability comes naturally from the inherent data and training process randomness, and also from the intrinsic non-identifiability of the model.

2.5.1   The Basic Ensemble Method

A simple approach to combining network outputs is to simply average them together. The basic ensemble method (BEM) output is defined:

$$f_{BEM} = \frac{1}{n} \sum_{i=1}^{n} f_i(x) \qquad (2.24)$$

This approach by itself can lead to improved performance, but doesn't take into account the fact that some flexible neural tree model may be more accurate than others. It has the advantage that it is easy to understand and implement and can be shown not to increase the expected error.

2.5.2   The Generalized Ensemble Method

A generalization to the BEM method is to find weights for each output that minimize the positive and negative classification rates of the ensemble. The general ensemble method (GEM) is defined as:

$$f_{BEM} = \sum_{i=1}^{n} \alpha_i f_i(x) \qquad (2.25)$$

where the $\alpha_i's$ are chosen to minimize the root mean square error between the Flexible neural tree model outputs and the desired values. For comparison purpose, the optimal weights of the ensemble predictor are optimized by using PSO algorithm.

2.5.3   The LWPR Method

To investigate more efficient ensemble method, a LWPR approximation approach is employed in this work[176]. In this framework, the final output of flexible neural tree ensemble is approximated by a local polynomial model, given by:

$$f_{LWPR} = \sum_{i=1}^{M} \beta_i t_i(x) \qquad (2.26)$$

where t_i is a function that produces the ith term in the polynomial. For example, with two inputs and a quadratic local model we would have $t_1(x) = 1$, $t_2(x) = x_1$, $t_3(x) = x_2$, $t_4(x) = x_1^2$, $t_5(x) = x_1 x_2$, $t_6(x) = x_2^2$. Equation (5) can be written more compactly as:

$$f_{LWPR} = \beta^T t(x) \tag{2.27}$$

where $t(x)$ is the vector of polynomial terms of the input x and β is the
vector of weight terms. The weight of the ith datapoint is computed as a
decaying function of Euclidean distance between x_k and x_{query}. β is chosen
to minimize:

$$\sum_{i=1}^{N} \omega_i^2 (f_{LWPR} - \beta^T t(x)) \tag{2.28}$$

where ω_i is a Gaussian weight function with kernel width K:

$$\omega_i = exp(-Distance^2(x_i, x_{query})/2K^2). \tag{2.29}$$

For this problem, an algorithm based on a multiresolution search of a quickly
constructible augmented kdtree without needing to rebuild the tree, has been
proposed for fast predictions with arbitrary local weighting functions [176].

2.5.4 Stock Index Forecasting Problem

We analyzed the seemingly chaotic behavior of two well-known stock indices namely the Nasdaq-100 index of NasdaqSM [177] and the S&P CNX
NIFTY stock index [178]. The Nasdaq-100 index reflects Nasdaq's largest
companies across major industry groups, including computer hardware and
software, telecommunications, retail/wholesale trade and biotechnology [177].
The Nasdaq-100 index is a modified capitalization-weighted index, which is
designed to limit domination of the Index by a few large stocks while generally
retaining the capitalization ranking of companies. Through an investment in
the Nasdaq-100 index tracking stock, investors can participate in the collective performance of many of the Nasdaq stocks that are often in the news or
have become household names. Similarly, S&P CNX NIFTY is a welldiversi-
fied 50 stock index accounting for 25 sectors of the economy [178]. It is used
for a variety of purposes such as benchmarking fund portfolios, index-based
derivatives and index funds. The CNX Indices are computed using market
capitalization weighted method, wherein the level of the Index reflects the
total market value of all the stocks in the index relative to a particular base
period. The method also takes into account constituent changes in the index
and importantly corporate actions such as stock splits, rights, etc. without
affecting the index value.

Prediction of stocks is generally believed to be a very difficult taskit be-
haves like a random walk process and time varying. The obvious complexity
of the problem paves the way for the importance of intelligent prediction
paradigms. During the last decade, stocks and futures traders have come
to rely upon various types of intelligent systems to make trading decisions

[161][162][163][164][168][172][179][171]. Several intelligent systems have in recent years been developed for modeling expertise, decision support and complicated automation tasks [174][180][175][185].

Leigh et al. [172] introduced a method for combining template matching, using pattern recognition and a feedforward neural network, to forecast stock market activity. The authors evaluated the effectiveness of the method for forecasting increases in the New York Stock Exchange Composite Index at a 5 trading day horizon. Results indicate that the technique is capable of returning results that are superior to those attained by random choice.

Kim and Chun [170] explored a new architecture for graded forecasting using an arrayed probabilistic network (APN) and used a "mistake chart" to compare the accuracy of learning systems against default performance based on a constant prediction. Authors also evaluated several backpropagation models against a recurrent neural network (RNN) as well as probabilistic neural networks, etc.

Tsaih et al. [183] investigated a hybrid AI (artificial intelligence) approach to the implementation of trading strategies in the S&P 500 stock index futures market. The hybrid AI approach integrates the rule-based systems technique and the neural networks technique to accurately predict the direction of daily price changes in S&P 500 stock index futures. By highlighting the advantages and overcoming the limitations of both the neural networks technique and rule-based systems technique, the hybrid approach can facilitate the development of more reliable intelligent systems to model expert thinking and to support the decision-making processes.

Refenes et al. [181] proposed a simple modification to the backpropagation procedure, which takes into account gradually changing inputCoutput relations. The procedure is based on the principle of discounted least squares whereby learning is biased towards more recent observations with long term effects experiencing exponential decay through time. This is particularly important in systems in which the structural relationship between input and response vectors changes gradually over time but certain elements of long term memory are still retained. The procedure is implemented by a simple modification of the least-squares cost function commonly used in error backpropagation.

Van den Berg et al. [184] proposed a probabilistic fuzzy systems to develop financial models where one can identify different states of the market for modifying ones actions. Authors developed a Takagi-Sugeno (TS) probabilistic fuzzy systems that combine interpretability of fuzzy systems with the statistical properties of probabilistic systems. The methodology is applied to financial time series analysis and demonstrated how a probabilistic TS fuzzy system can be identified, assuming that a linguistic term set is given.

From the perspective of the agent-based model of stock markets, Chen and Liao [165] examined the possible explanations for the presence of the causal relation between stock returns and trading volume. Using the agent-based approach, the authors found that the explanation for the presence of the stock price volume relation may be more fundamental. Conventional devices such as information asymmetry, reaction asymmetry, noise traders or tax motives are not explicitly required. Authors claimed that a full understanding of the price volume relation cannot be accomplished unless the feedback relation between individual behavior at the bottom and aggregate phenomena at the top is well understood.

We investigated the performance analysis of flexible neural tree [189][190] [166] ensemble for modeling the Nasdaq-100 and the NIFTY stock market indices. The hierarchical structure of flexible neural tree is evolved using genetic programming with specific instructions. The parameters of the flexible neural tree model are optimized by particle swarm optimization algorithm [136]. The method interleaves both optimizations. Starting with random structures and corresponding parameters, it first tries to improve the structure and then as soon as an improved structure is found, it fine tunes its parameters. It then goes back to improving the structure again and, fine tunes the structure and rules parameters. This loop continues until a satisfactory solution is found or a time limit is reached.

2.6 Stock Index Forecasting Experimental Illustrations

We analyzed the Nasdaq-100 index value from 11 January 1995 to 11 January 2002 [177] and the NIFTY index from 01 January 1998 to 03 December 2001 [178]. For both the indices, we divided the entire data into almost two equal parts. No special rules were used to select the training set other than ensuring a reasonable representation of the parameter space of the problem domain [163].

We considered 7-year stock data for the Nasdaq-100 Index and 4-year for the NIFTY index. Our target is to develop efficient forecast models that could predict the index value of the following trade day based on the opening, closing and maximum values of the same on a given day. The assessment of the prediction performance of the different ensemble paradigms were done by quantifying the prediction obtained on an independent data set. The Root Mean Squared Error (RMSE), Maximum Absolute Percentage Error (MAP) and Mean Absolute Percentage Error (MAPE) and Correlation Coefficient (CC) were used to study the performance of the trained forecasting model for the test data. MAP is defined as follows:

$$MAP = max(\frac{|P_{actual,i} - P_{predicted,i}|}{P_{predicted,i}} \times 100) \qquad (2.30)$$

where $P_{actual,i}$ is the actual index value on day i and $P_{predicted,i}$ is the forecast value of the index on that day. Similarly $MAPE$ is given as:

Table 2.27 Empirical comparison of RMSE results for four learning methods

	Best-FNT	BEM	GEM	LWPR
Nasdaq-100	0.01854	0.01824	0.01635	4.41×10^{-5}
NIFTY	0.01315	0.01258	0.01222	1.96×10^{-7}

Table 2.28 Statistical analysis of four learning methods (test data)

	Best-FNT	BEM	GEM	LWPR
		Nasdaq-100		
CC	0.997542	0.997610	0.997757	0.999999
MAP	98.1298	98.3320	97.3347	0.4709
MAPE	6.1090	6.3370	5.7830	0.0040
		NIFTY		
CC	0.996908	0.997001	0.0997109	0.999999
MAP	28.0064	34.3687	26.8188	7.65×10^{-4}
MAPE	3.2049	2.9303	2.6570	1.92×10^{-5}

$$MAPE = \frac{1}{N} \sum_{i=1}^{N} \left(\frac{|P_{actual,i} - P_{predicted,i}|}{P_{predicted,i}} \right) \times 100 \qquad (2.31)$$

where N represents the total number of days.

We used instruction set $I = \{+_2, +_3, \ldots, +_6, x_0, x_1, x_2\}$ for modeling the Nasdaq-100 index and instruction set $I = \{+_2, +_3, \ldots, +_8, x_0, x_1, x_2, x_3, x_4\}$ for modeling the NIFTY index. We have conducted 10 flexible neural tree models for predicting the Nasdaq-100 index and the NIFTY index, respectively and the three ensemble methods discussed above are employed to predict both indices.

Table 2.27 summarizes the test results achieved for the two stock indices using the four different approaches. Performance analysis of the trained forecasting models for the test data is illustrated in Table 2.28. Figures 2.34 and 2.35 depict the test results for the one day ahead prediction of the Nasdaq−100 index and the NIFTY index respectively.

We have demonstrated how the chaotic behavior of stock indices could be well represented by FNT ensemble learning paradigm. Empirical results on the two data sets using FNT ensemble models clearly reveal the efficiency of the proposed techniques. In terms of RMSE values, for the Nasdaq-100 index and the NIFTY index, LWPR performed marginally better than other models. For both indices (test data), LWPR also has the highest correlation coefficient and the lowest value of MAPE and MAP values. A low MAP value

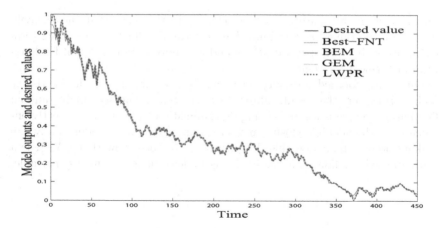

Fig. 2.34 Test results showing the performance of the different methods for modeling the Nasdaq-100 index

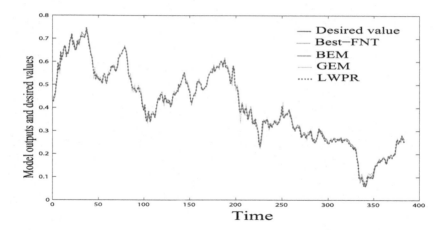

Fig. 2.35 Test results showing the performance of the different methods for modeling the NIFTY index

is a crucial indicator for evaluating the stability of a market under unforeseen fluctuations. In the present example, the predictability assures the fact that the decrease in trade off is only a temporary cyclic variation that is perfectly under control.

Models were built to predict the share price for the following trade day based on the opening, closing and maximum values of the same on a given day. The experimental results indicate that the most prominent parameters that affect share prices are their immediate opening and closing values. The fluctuations in the share market are chaotic in the sense that they heavily depend on the values of their immediate forerunning fluctuations. Long-term trends exist, but are slow variations and this information is useful for long-term investment strategies.

Our study focused on short term, on floor trades, in which the risk is higher. However, the results illustrate that even in the seemingly random fluctuations, there is an underlying deterministic feature that is directly enciphered in the opening, closing and maximum values of the index of any day making predictability possible. Empirical results also show that LWPR is a distinguished candidate for the FNT ensemble or neural networks ensemble.

Part III

Hierarchical Neural Networks

3

Hierarchical Neural Networks

Summary. Soft Computing (SC), including Neural Computing (NC), Fuzzy Computing (FC), Evolutionary Computing (EC) etc., provides us with a set of flexible computing tools to perform approximate reasoning, learning from data and search tasks. Moreover, it has been observed that the highly increasing computing power and technology, could make possible the use of more complex intelligent architectures, taking advantage of more than one intelligent techniques, not in a competitive, but rather in a collaborative sense. Therefore, discovering of more sophisticated and new evolutionary learning models and its application to new areas and problems still remain as key questions for the next 10 years.

There are three basic multilevel structures for hierarchical models, namely, incremental, aggregated and cascaded. Designing of these hierarchical models faces many difficulties including determination of the hierarchical structure, parameter identification and input variables selection for each sub-models.

This Chapter provides a gentle introduction to three different hierarchical architectures. First the design and implementation of hierarchical radial basis function networks are illustrated for breast cancer detection and face recognition. Further, the development of hierarchical B-spline networks is illustrated for breast cancer detection and time series prediction. Finally, hierarchical wavelet neural networks are presented for several function approximation problems.

3.1 Hierarchical Radial Basis Function Neural Networks

Hierarchical radial basis function networks (HRBF) consist of multiple RBF networks assembled in different levels or cascade architecture in which a problem was divided and solved in more than one step. Mat Isa et al. used Hierarchical Radial Basis Function (HiRBF) to increase RBF performance in

Y. Chen, A. Abraham.: Tree-Struc. Based Hybrid Com. Intelligence, ISRL 2, pp. 99–125.
springerlink.com © Springer-Verlag Berlin Heidelberg 2010

diagnosing cervical cancer [186]. The HiRBF cascaded together two RBF networks, where both networks have different structure but using the same learning algorithms. The first network classifies all data and performs a filtering process to ensure that only certain attributes to be fed to the second network. Their research illustrated that the HiRBF performs better compared to single RBF. Hierarchical RBF network has been proved effective in the reconstruction of smooth surfaces from sparse noisy data points [187]. In order to improve the model generalization performance, a selective combination of multiple neural networks by using Bayesian method was proposed in [188].

For real engineering andor scientific problems, how to automatically design a hierarchical RBF network remains unsolved. Based on flexible tree representation and evolutionary algorithms, we illustrate an optimal design method for the HRBF automatically.

3.1.1 The Radial Basis Function Network

A radial basis function (RBF) network is a feed-forward neural network with one hidden layer of RBF units and a linear output layer. By an RBF unit we mean a neuron with multiple real inputs $\mathbf{x} = (x_1, \ldots, x_n)$ and one output y computed as:

$$y = \varphi(\xi); \quad \xi = \frac{\|\mathbf{x} - \mathbf{c}\|_C}{b} \tag{3.1}$$

where $\varphi : R \to R$ is a suitable activation function, let us consider Gaussian radial basis function $\varphi(z) = e^{-z^2}$. The center $\mathbf{c} \in R^n$, the width $b \in R$ and an $n \times n$ real matrix $\mathbf{C}$ are a unit's parameters, $\| \cdot \|_C$ denotes a weighted norm defined as $\|\mathbf{x}\|_C^2 = (\mathbf{Cx})^T(\mathbf{Cx}) = \mathbf{x}^T \mathbf{C}^T \mathbf{Cx}$.

Thus, the network represents the following real function $\mathbf{f} : R^n \to R^m$:

$$f_s(\mathbf{x}) = \sum_{j=1}^{h} w_{js} e^{-\left(\frac{\|\mathbf{x} - \mathbf{c}\|_C}{b}\right)^2}, \qquad s = 1, \ldots, m, \tag{3.2}$$

where $w_{js} \in R$ are weights of s-th output unit and f_s is the s-th network output.

The goal of an RBF network learning is to find suitable values of RBF units' parameters and the output layer's weights, so that the RBF network function approximates a function given by a set of examples of inputs and desired outputs $T = \{\mathbf{x}(t), \mathbf{d}(t); t = 1, \ldots, k\}$, called a *training set*. The quality of the learned RBF network is usually measured by the *error function*:

$$E = \frac{1}{2} \sum_{t=1}^{k} \sum_{j=1}^{m} e_j^2(t), \qquad e_j(t) = d_j(t) - f_j(t). \tag{3.3}$$

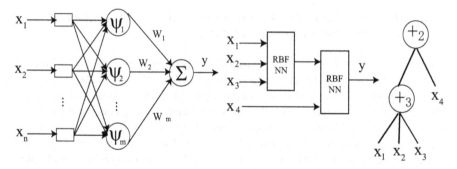

Fig. 3.1 A RBF neural network (left), an example of hierarchical RBF network (middle), and a tree-structural representation of the HRBF network (right)

3.1.2 Automatic Design of Hierarchical Radial Basis Function Network

Encoding and Calculation for Hierarchical Radial Basis Function Network

A function set F and terminal instruction set T used for generating a HRBF network model are described as $S = F \bigcup T = \{+_2, +_3, \ldots, +_N\} \bigcup \{x_1, \ldots, x_n\}$, where $+_i (i = 2, 3, \ldots, N)$ denote non-leaf nodes' instructions and taking i arguments. $x_1, x_2, \ldots, x_n$ are leaf nodes' instructions and taking no arguments. The output of a non-leaf node is calculated as a HRBF network model (see Figure 3.1). From this point of view, the instruction $+_i$ is also called a basis function operator with i inputs.

Gaussian radial basis function is used and the number of radial basis functions used in hidden layer of the network is the same as the number of inputs, that is, $m = n$.

In the construction process of a HRBF network tree, if a nonterminal instruction, i.e., $+_i (i = 2, 3, 4, \ldots, N)$ is selected, i real values are randomly generated and used for representing the connection strength between the node $+_i$ and its children. In addition, $2 \times n^2$ adjustable parameters a_i and b_i are randomly created as radial basis function parameters. The output of the node $+_i$ can be calculated by using Eqn. 3.1 and Eqn. 3.2. The overall output of HRBF network tree can be computed from left to right by depth-first method, recursively.

Finding an optimal or near-optimal HRBF network structure is formulated as a product of evolution. Genetic Programming (GP) and Probabilistic Incremental Program Evolution (PIPE) have been widely explored for structure optimization of the FNT [189][190]. We illustrate the usage of Extended Compact Genetic Programming (ECGP) [191] to find an optimal or near-optimal structure of HRBF networks.

3.1.3   Tree Structure Optimization by Extended Compact Genetic Programming (ECGP)

Finding an optimal or near-optimal HRBF is formulated as a product of evolution. ECGP is a direct extension of ECGA to the tree representation, which is based on the PIPE prototype tree. In ECGA, Marginal Product Models (MPMs) are used to model the interaction among genes, represented as random variables, given a population of genetic algorithm individuals. MPMs are represented as measures of marginal distributions on partitions of random variables. ECGP is based on the PIPE prototype tree, and thus each node in the prototype tree is a random variable. ECGP decomposes or partitions the prototype tree into sub-trees, and the MPM factorises the joint probability of all nodes of the prototype tree, to a product of marginal distributions on a partition of its sub-trees. A greedy search heuristic is used to find an optimal MPM mode under the framework of minimum encoding inference. ECGP can represent the probability distribution for more than one node at a time. Thus, it extends PIPE in that the interactions among multiple nodes are considered.

3.1.4   Parameter Optimization Using Differential Evolution Algorithm

Differential Evolution (DE) algorithm was first introduced by Storn and Price in 1995 [192]. It resembles the structure of an evolutionary algorithm (EA), but differs from traditional EAs in its generation of new candidate solutions and by its use of a 'greedy' selection scheme. DE works as follows: First, all individuals are randomly initialized and evaluated using the fitness function provided. Afterwards, the following process will be executed as long as the termination condition is not fulfilled: For each individual in the population, an offspring is created using the weighted difference of parent solutions. The offspring replaces the parent if it is fitter. Otherwise, the parent survives and is passed on to the next iteration of the algorithm. In generation k, we denote the population members by x_1^k, x_2^k, ..., x_N^k. The DE algorithm is given as follows [193]:

S1 Set $k = 0$, and randomly generate N points x_1^0, x_2^0, ..., x_N^0 from search space to form an initial population;

S2 For each point $x_i^k (1 \leq i \leq N)$, execute the DE offspring generation scheme to generate an offspring $x_i^{(k+1)}$;

S3 If the given stop criteria is not met, set $k = k + 1$, goto step S2.

The DE offspring generation approach used is given as follows:

S1 Choose one point x_d randomly such that $f(x_d)$ $f(x_i^k)$, another two points x_b, x_c randomly from the current population and a subset

$S = \{j_1, \ldots, j_m\}$ of the index set $\{1, \ldots, n\}$, while $m < n$ and all j_i mutually different;

S2 Generate a trial point $u = (u_1, u_2, \ldots, u_n)$ as follows:

DE Mutation. Generate a temporary point z as follows,

$$z = (F + 0.5)x_d + (F - 0.5)x_i + F(x_b - x_c); \qquad (3.4)$$

Where F is a give control parameter;

DE Crossover. For $j \in S$, u_j is chosen to be z_j; otherwise u_j is chosen a to be $(x_i^k)_j$;

S3 If $f(u) \le f(x_i^k)$, set $x_i^{k+1} = u$; otherwise, set $x_i^{k+1} = x_i^k$.

3.1.5 Procedure of The General Learning Algorithm

The general learning procedure for constructing the HRBF network can be described as follows.

S1 Create an initial population randomly (HRBF network trees and its corresponding parameters);

S2 Structure optimization is achieved by using ECGP algorithm;

S3 If a better structure is found, then go to step S4, otherwise go to step S2;

S4 Parameter optimization is achieved by DE algorithm. In this stage, the architecture of HRBF network model is fixed, and it is the best tree developed during the end of run of the structure search;

S5 If the maximum number of local search is reached, or no better parameter vector is found for a significantly long time then go to step S6; otherwise go to step S4;

S6 If satisfactory solution is found, then the algorithm is stopped; otherwise go to step S2.

3.1.6 Variable Selection in the HRBF Network Paradigms

It is often a difficult task to select important variables for a classification or regression problem, especially when the feature space is large. Conventional RBF neural network usually cannot do this. In the perspective of HRBF network framework, the nature of model construction procedure allows the HRBF network to identify important input features in building a HRBF network model that is computationally efficient and effective. The mechanisms of input selection in the HRBF network constructing procedure are as follows:

- Initially the input variables are selected to formulate the HRBF network model with same probabilities;

- The variables which have more contribution to the objective function will be enhanced and have high opportunity to survive in the next generation by an evolutionary procedure;
- The evolutionary operators i.e., crossover and mutation, provide a input selection method by which the HRBF network should select appropriate variables automatically.

3.1.7 Experimental Illustrations

Wisconsin Breast Cancer Detection

We made use of the Wisconsin breast cancer data set from the UCI machine-learning database repository [196]. This data set has 30 attributes (30 real valued input features) and 569 instances of which 357 are of benign and 212 are of malignant type. The data set is randomly divided into a training data set and a test data set. The first 285 data is used for training and the remaining 284 data is used for testing the performance of the different models.

All the models were trained and tested with the same set of data. The instruction sets used to create an optimal HRBF network classifier is $S = F \bigcup T = \{+_2, \ldots, +_5\} \bigcup \{x_0, x_1, \ldots, x_{29}\}$, where $x_i (i = 0, 1, \ldots, 29)$ denotes the 30 input features. The optimal hierarchical HRBF network for breast cancel detection problem is shown in Figure 3.2. The classification results for testing data set are shown in Table 3.1. For comparison purpose, the detection performances of the FNT, NN and RBF-NN are also shown in Table 3.1 (for details, see [167]). The important features for constructing the HRBF network models are shown in Table 3.2. It should be noted that the obtained HRBF network classifier has smaller size and reduced features and with high accuracy for breast cancer detection. Receiver Operating Characteristics (ROC) analysis of the FNT, NN, RBF-NN and the HRBF network model is shown in Table 3.3.

Table 3.1 Comparative results of the FNT, NN, RBF [194] and the proposed HRBF network classification methods for the detection of breast cancer

Cancer type	FNT(%)	NN(%)	RBF-NN(%)	HRBF(%)
Benign	93.31	94.01	94.12	96.83
Malignant	93.45	95.42	93.21	96.83

Table 3.2 The important features selected by the HRBF network

$$x_0, \ x_1, \ x_2, \ x_3, \ x_6, \ x_7, \ x_9, \ x_{18}, \ x_{20}, \ x_{25}, \ x_{27}, \ x_{29}$$

Table 3.3 Comparison of false positive rate (fp) and true positive rate (tp) for FNT, NN, RBF-NN [234] and hierarchical HRBF network

Cancer Type	FNT fp(%)	tp(%)	NN fp(%)	tp(%)	RBF-NN fp(%)	tp(%)	HRBF fp(%)	tp(%)
Benign	3.88	91.71	4.85	93.37	6.6	97.14	2.91	96.69
Malignant	2.76	86.41	4.97	96.12	9.2	96.87	3.31	97.09

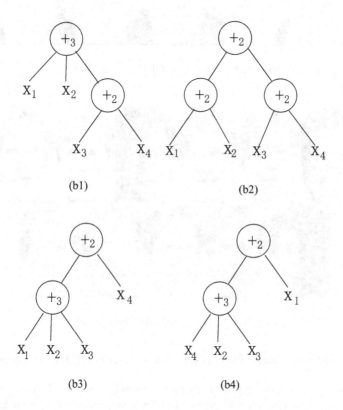

Fig. 3.2 The optimized HRBF network for breast cancer detection

3.1.8   Face Recognition

HRBF network is employed for face recognition and we used Discrete Cosine Transform (DCT) for feature selection, which is the same as discussed in the last Chapter.

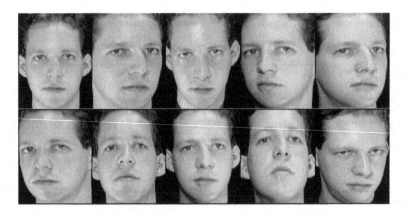

Fig. 3.3 Example in ORL face dataset

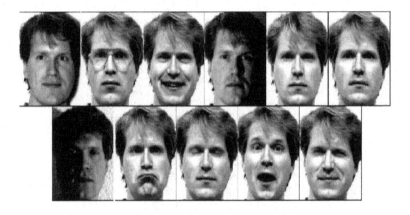

Fig. 3.4 Example in Yale face dataset

Feature Selection and Face recognition Using HRBF Paradigms

We performed extensive experiments on two benchmark face datasets, namely the ORL and the Yale face database. In all the experiments, the background is cut out, and the images are resized to 92×112. No other preprocessing is done. Besides HRBF, the PCA based method, LDA-based method, neural networks etc. were also tested for comparisons.

The Face Database

For ORL face dateset, 40 persons with variations in facial expression (e.g. open/close eyes, smiling/non-smiling), and facial details (e.g. wearing glasses/ not wearing glasses). All images were taken under a dark background, and the subjects were in an upright frontal position, with tilting and rotation tolerance up to 20 degree, and tolerance of up to about 10%. For each person, 5 images are generated randomly to form the training data set and the remaining were chosen as test data set. Figure 3.3 shows 12 images of one subject from the selected dataset.

The Yale face database contains 165 images of 15 subjects. There are 11 images per subject with different facial expressions or lightings. Figure 3.4 shows the 11 images of one subject. For each person, 5 images are generated randomly to form the training data set and the remaining were chosen as test data set.

Experiments Using ORL and Yale Face Database

For this simulation, the DCT is employed to training and test data sets, respectively. The extracted 60 input features are used for constructing a HRBF model. A HRBF classifier was constructed using the training data and then the classifier was used on the test data set to classify the data as a face ID or not. The instruction sets used to create an optimal HRBF classifier is $S = \{+_2, +_3, \ldots, +_6, x_0, x_1, \ldots, x_{59}\}$, where $x_i (i = 0, 1, \ldots, 59)$ denotes the 60 features extracted by DCT.

A comparison of different feature extraction methods and different face classification methods for ORL face dataset is shown in Table 3.4. Table 3.5 depicts the face recognition performance of the HRBF by using the 60 features for Yale data set. The HRBF method helps to reduce the features from 60 to 6-15.

Facial features were first extracted by the DCT, which greatly reduces dimensionality of the original face image as well as maintains the main facial features. Compared with the well-known PCA approach, the DCT has the advantages of data independency and fast computational speed. The presented HRBF model for face recognition was focused on improving the face recognition performance by reducing the input features. Simulation results on ORL and Yale face database also illustrate that the HRBF method achieves

Table 3.4 Comparison of different approaches for ORL face recognition (test)

Method	Recognition rate
PCA+RBF [197]	94.5%
LDA+RBF [197]	94.0%
FS+RBF [197]	92.0%
NN [198]	94.64%
PCA [198]	88.31%
LDA [198]	88.87%
DCT+HRBF	97.89%

Table 3.5 Comparison of different approaches for Yale face recognition (test)

Method	Recognition rate
NN [197]	83.51%
PCA [197]	81.13%
LDA [197]	98.69%
DCT+HRBF	99.25%

high training and recognition speed, as well as high recognition rate. More importantly, it is insensitive to illumination variations.

3.2　Hierarchical B-Spline Neural Networks

Hierarchical B-spline networks consist of multiple B-spline networks assembled in different level or cascade architecture. We illustrate an automatic method for constructing hierarchical B-spline network. Based on a pre-defined instruction/operator set, the hierarchical B-spline networks can be created and evolved. The hierarchical B-spline network allows input variables selection. The hierarchical structure is evolved using the Extended Compact Genetic Programming (ECGP), a tree-structure based evolutionary algorithm. The fine tuning of the parameters encoded in the structure is accomplished using Particle Swarm Optimization (PSO). The proposed method interleaves both optimizations.

3.2.1　The B-Spline Network

B-spline basis functions are piecewise polynomials, producing models with a response of a desired smoothness. The order of these local polynomials is denoted by the order of the B-spline, denoted by k. A set of univariate basis functions is denoted on a series of knots, which represent the piecewise polynomial intervals. A knot vector for a set of order k univariate basis functions is described by, $\lambda = (\lambda_0, \lambda_1, \ldots, \lambda_{k+r-1})^T$, where r is the number of univariate basis functions denoted on this variable and λ_j is the jth knot. The input domain of a set of univariate basis functions is given by $X = [\lambda_{k-1}, \lambda_k]$ giving a total of $(r - k)$ interior knots. The exterior knots are those that lie outside this domain, not significantly affecting the modeling capabilities of the model. But they do influence the numerical conditioning of the weight optimization problem.

Univariate B-spline basis functions are calculated using a numerical stable computationally efficient recurrence relationship given by:

$$N_k^j(x) = \frac{x - \lambda_{j-k}}{\lambda_{j-1} - \lambda_{j-k}} N_{k-1}^{j-1}(x) + \frac{\lambda_j - x}{\lambda_j - \lambda_{j-k+1}} N_{k-1}^j(x) \qquad (3.5)$$

$$N_1^j(x) = \begin{cases} 1, \, x \in [\lambda_{j-1}, \lambda_j] \\ 0, \, otherwise \end{cases} \qquad (3.6)$$

where $N_k^j(\cdot)$ is the jth univariate basis function of order k.

To define multivariate B-spline basis functions, the tensor product of the univariate basis functions is performed. Given a set of B-spline basis functions defined across each input variable x_j of a specified order k_j, a multivariate B-spline basis function is calculated by multiplying the n univariate membership functions $N_{k_j}^{j_i}(x_j)$ giving:

$$N_\mathbf{k}^i(\mathbf{x}) = \prod_{j=1}^n N_{k_j}^{j_i}(x_j) \qquad (3.7)$$

where j_i represents the index to the basis function, defined on x_j, which contributes to the ith multivariate basis function.

A B-spline Neural Network is composed of three layers, which are, a normalized input layer, a basis functions layer and a linear weight layer (see Figure 3.5). In general, the B-spline basis function network can be represented as:

$$y = \sum_{i=1}^m \omega_i \psi_i(x; \theta) \qquad (3.8)$$

where $x \in R^n$ is input vector, $\psi_i(x; \theta)$ is ith B-spline basis function, and ω_i is the corresponding weights of ith basis function and θ is the parameter vector used in the basis functions.

3.3 Automatic Design of HB-Spline Network

3.3.1 Encode and Calculation for HB-Spline

A function set F and terminal instruction set T used for generating a hierarchical B-spline network model are described as $S = F \bigcup T = \{+_2, +_3, \ldots, +_N\}$ $\bigcup \{x_1, \ldots, x_n\}$, where $+_i (i = 2, 3, \ldots, N)$ denote non-leaf nodes' instructions and taking i arguments. $x_1, x_2, \ldots, x_n$ are leaf nodes' instructions and taking no arguments. The output of a non-leaf node is calculated as a B-spline neural network model (see Figure 3.5). From this point of view, the instruction $+_i$ is also called a basis function operator with i inputs.

The translation and dilation of order 3 B-spline function is used as basis function, for experimental illustrations,

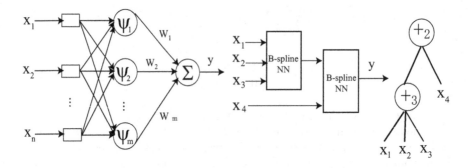

Fig. 3.5 A B-spline network (left), a hierarchical B-spline network (middle), and a tree-structural representation of the B-spline network (right)

$$N_3(a,b,x) = \begin{cases} \frac{9}{8} + \frac{3}{2}(\frac{x-b}{a}) + \frac{1}{2}(\frac{x-b}{a})^2, & x \in [-\frac{3}{2}a + b, -\frac{1}{2}a + b) \\ \frac{3}{4} - (\frac{x-b}{a})^2, & x \in [-\frac{1}{2}a + b, \frac{1}{2}a + b) \\ \frac{9}{8} - \frac{3}{2}(\frac{x-b}{a}) + \frac{1}{2}(\frac{x-b}{a})^2, & x \in [\frac{1}{2}a + b, \frac{3}{2}a + b] \\ 0, & otherwise \end{cases} \quad (3.9)$$

and the number of B-spline basis functions used in hidden layer of the network is same with the number of inputs, that is, $m = n$.

In the construction process of hierarchical B-spline network tree, if a non-terminal instruction, i.e., $+_i(i = 2, 3, 4, \ldots, N)$ is selected, i real values are randomly generated and used for representing the connection strength between the node $+_i$ and its children. In addition, $2 \times n^2$ adjustable parameters a_i and b_i are randomly created as B-spline basis function parameters. The output of the node $+_i$ can be calculated by using Eqn. 3.7 and Eqn. 3.8. The overall output of hierarchical B-spline network tree can be computed from left to right by depth-first method, recursively.

3.3.2 Tree Structure and Parameter Optimization

Finding an optimal or near-optimal hierarchical B-spline network structure is formulated as a product of evolution. Any kind of tree structure based evolutionary algorithms, i.e., the Genetic Programming (GP), Probabilistic Incremental Program Evolution (PIPE), Gene Expression Programming (GEP), Ant Programming (AP), Extended Compact Genetic Programming (ECGP) etc. can be used to find optimal structure of the HB-Spline network. Particle swarm optimization (PSO) [136] algorithm is used for parameter optimization for experimental illustrations.

3.3.3 Procedure of the General Learning Algorithm

The general learning procedure for constructing the hierarchical B-spline network is summarized as follows:

S1 Create an initial population randomly (hierarchical B-spline trees and its corresponding parameters);

S2 Structure optimization is achieved by using tree-structure based evolutionary algorithm;

S3 If a better structure is found, then go to step S4, otherwise go to step S2;

S4 Parameter optimization is achieved by PSO algorithm. In this stage, the architecture of hierarchical B-spline network model is fixed, and it is the best tree developed during the end of run of the structure search;

S5 If the maximum number of local search is reached, or no better parameter vector is found for a significantly long time then go to step S6; otherwise go to step S4;

S6 If satisfactory solution is found, then the algorithm is stopped; otherwise go to step S2.

3.3.4 Variable Selection in the Hierarchical B-Spline Network Paradigms

It is often a difficult task to select important variables for a classification or regression problem, especially when the feature space is large. Conventional B-spline neural network usually cannot do this. In the perspective of hierarchical B-spline framework, the nature of model construction procedure allows the hierarchical B-spline network to identify important input features in building a hierarchical B-spline network model that is computationally efficient and effective. The mechanisms of input selection in the hierarchical B-spline network constructing procedure are as follows.

- Initially the input variables are selected to formulate the hierarchical B-spline network model with same probabilities;
- The variables which have more contribution to the objective function will be enhanced and have high opportunity to survive in the next generation by an evolutionary procedure;
- The evolutionary operators i.e., crossover and mutation, provide a input selection method by which the hierarchical B-spline network should select appropriate variables automatically.

3.3.5 Experimental Illustrations

3.3.6 Wisconsin Breast Cancer Detection

We used the Wisconsin breast cancer data set from the UCI machine-learning database repository [196]. This data set has 30 attributes (30 real valued

Table 3.6 Comparative results of the FNT, NN, WNN [167] and the proposed hierarchical B-spline network classification methods for the detection of breast cancer

Cancer type	FNT(%)	NN(%)	WNN(%)	H-Bspline(%)
Benign	93.31	94.01	94.37	96.77
Malignant	93.45	95.42	92.96	96.77

Table 3.7 The important features selected by the hierarchical B-spline network

$$x_0, \ x_2, \ x_3, \ x_7, \ x_9, \ x_{18}, \ x_{21}$$

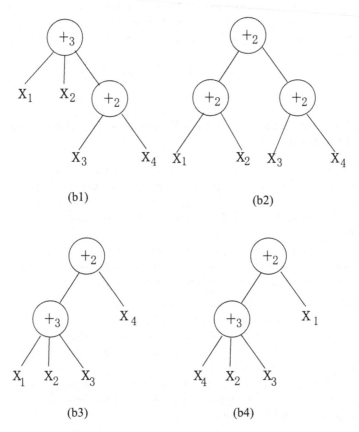

Fig. 3.6 The optimized hierarchical B-spline network for breast cancel detection

input features) and 569 instances of which 357 are of benign and 212 are of malignant type. The data set is randomly divided into a training data set and a test data set. The first 285 data is used for training and the remaining 284 data is used for testing the performance of the different models.

Table 3.8 Comparison of false positive rate (fp) and true positive rate (tp) for FNT, NN, WNN [167] and hierarchical B-spline network

Cancer Type	FNT		NN		WNN		H-Bspline	
	fp(%)	tp(%)	fp(%)	tp(%)	fp(%)	tp(%)	fp(%)	tp(%)
Benign	3.88	91.71	4.85	93.37	6.8	98.34	4.85	96.13
Malignant	2.76	86.41	4.97	96.12	9.4	97.09	3.87	95.15

All the models were trained and tested with the same set of data. The instruction sets used to create an optimal hierarchical B-spline network classifier is:

$$S = F \bigcup T = \{+_2, +_3, \ldots, +_5\} \bigcup \{x_0, x_1, \ldots, x_{29}\}. \tag{3.10}$$

Where $x_i (i = 0, 1, \ldots, 29)$ denotes the 30 input features. The optimal hierarchical B-spline network for breast cancer detection problem is shown in Figure 3.6. The classification results for testing data set are shown in Table 3.6. For comparison purpose, the detection performances of the FNT, NN and WNN are also depicted in Table 3.6 (for details, see [167]). The important features for constructing the hierarchical B-spline models are shown in Table 3.7. It should be noted that the obtained hierarchical B-spline network classifier has smaller size, reduced features and with high accuracy for breast cancer detection. Receiver Operating Characteristics (ROC) analysis of the FNT, NN, WNN and the hierarchical B-spline network model is illustrated in Table 3.8.

We illustrated the automatic design of an optimized hierarchical B-spline network for the detection of breast cancer and compared the results with some advanced artificial intelligence techniques, i.e., FNT, NN and Wavelet neural network (WNN). As depicted in Table 3.6, the preliminary results are very encouraging. The best accuracy was offered by the hierarchical B-spline network method followed by the wavelet neural network for detecting benign types and PSO trained neural network for detecting the malignant type of cancer. ROC analysis (Table 3.8) illustrates that wavelet neural network has the highest false positive rate and the H-Bspline and FNT models have the lowest false positive rates for detecting benign and malignant cancer, respectively.

3.3.7 Time-Series Forecasting

Developed hierarchical B-spline network models are applied for two time-series prediction problems: Wolfer sunspots series and Box-Jenkins time

Table 3.9 Parameters Used In Hierarchical B-spline Network Model

Parameter	Initial value
Population Size PS	30
Elitist Learning Probability P_{el}	0.01
Learning rate lr	0.01
Fitness constant ε	0.000001
Overall mutation probability P_M	0.4
Mutation rate mr	0.4
Prune threshold T_P	0.999999
Maximum local search steps	1000
Initial connection weights	rand[-1, 1]
Initial parameters a_i and b_i	rand[0,1]

series. Well-known benchmark examples are used for the sake of an easy comparison with existing models. For the structure optimization, PIPE algorithm is used. The parameters used for each experiment are listed in Table 3.9.

Wolfer Sunspots Series Prediction

For the time series prediction task, 100 observations of the Wolfer sunspots series were used with an embedding dimension of 10 and a delay time of one. This data is a time series of the annual Wolfer Sunspot average number of sunspots on the sun during each year. The data is normalized in the range [0 1] for experimentation. 80 data samples were used for training and other 10 for testing.

The used instruction sets to create an optimal hierarchical B-spline network model is $S = F \bigcup T = \{+_2, +_3, +_4\} \bigcup \{x_0, x_1, \ldots, x_9\}$ where $x_i (i = 0, 1, \ldots, 9)$ denotes $x(t)$, $x(t-1)$, $\ldots$, $x(t-9)$, respectively.

After 15 generations of the evolution, the optimal hierarchical B-spline network model was obtained with MSE 0.0019. The MSE value for validation data set is 0.0012. The evolved hierarchical B-spline network tree is shown in Figure 3.7 and the actual time-series, the hierarchical B-spline network model output is shown in Figure 3.8. From the evolved hierarchical B-spline network tree, it can be seen that the optimal inputs variables for constructing a hierarchical B-spline tree model are: $x(t)$, $x(t-2)$, $x(t-3)$, $x(t-6)$, $x(t-7)$ and $x(t-8)$. For comparison purpose, a B-spline network and a feedforward neural network is also trained to predict the Wolfer sunspot time-series. A comparison result of different methods for forecasting Wolfer sunspot time-series is shown in Table 3.10. It should be noted that the hierarchical B-spline network model with proper selected input variables has better precision and good generalization ability.

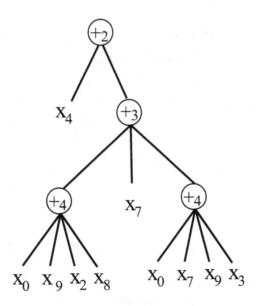

Fig. 3.7 The evolved structure of hierarchical B-spline network model for prediction of Wolfer sunspots series

Table 3.10 Comparison of prediction errors using different methods for the Wolfer sunspots series

Model name	Number of inputs	MSE
ANN model	10	0.0071
B-spline network)	10	0.0066
HB-Spline network	7	0.0012

Application to Jenkins-Box Time-Series

The gas furnace data (series J) of Box and Jenkins (1970) was recorded from a combustion process of a methane-air mixture. It is well known and frequently used as a benchmark example for testing identification and prediction algorithms. The data set consists of 296 pairs of input-output measurements. The input $u(t)$ is the gas flow into the furnace and the output $y(t)$ is the CO_2 concentration in outlet gas. The sampling interval is 9s. For illustration, 10 inputs variables are used for constructing a hierarchical B-spline network model. The appropriate time-lags for constructing a hierarchical B-spline model are finally determined by an evolutionary procedure.

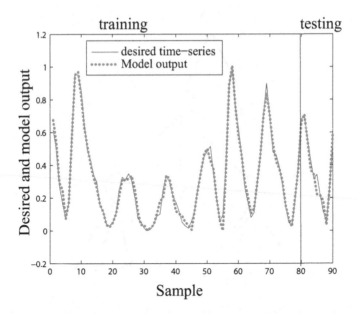

Fig. 3.8 The actual time series data, output of the evolved hierarchical B-spline network model and the prediction error for training and test samples

The used instruction sets to create an optimal hierarchical B-spline model is $S = F \bigcup T = \{+_2, +_3\} \bigcup \{x_0, x_1, \ldots, x_9\}$, where $x_i(i = 0, 1, \ldots, 9)$ denotes $y(t-1)$, $y(t-2)$, $y(t-3)$, $y(t-4)$, and $u(t-1)$, $u(t-2)$, $u(t-3)$, $u(t-4)$, $u(t-5)$ and $u(t-6)$, respectively.

After 21 generations of the evolution, the optimal hierarchical b-spline network model was obtained with MSE 0.00108. The MSE value for validation data set is 0.00123. The evolved hierarchical B-spline network tree is shown in Figure 3.9, and the actual time-series, the hierarchical B-spline network model output is depicted in Figure 3.10. From the evolved hierarchical B-spline network tree, it is evident that the optimal inputs variables for constructing a hierarchical B-spline network model are: $u(t-4)$, $u(t-5)$, $u(t-6)$, $y(t-1)$ and $y(t-3)$. It should be noted that the hierarchical B-spline network model with proper selected input variables has better precision and good generalization ability. A comparison result of different methods for forecasting Jenkins-Box data is shown in Table 3.11.

It is evident that the hierarchical B-spline network model works well for generating prediction models of time series. Preliminary research results reveal that the evolved hierarchical B-spline network models are effective for time-series prediction problems.

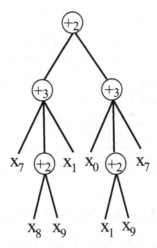

Fig. 3.9 The evolved structure of hierarchical B-spline network model for prediction of Jenkins-Box data

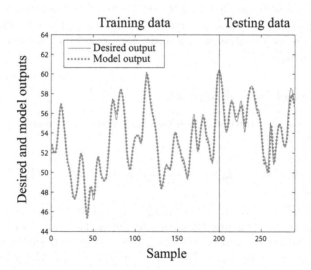

Fig. 3.10 The actual time series data, output of the evolved hierarchical B-spline network model and the prediction error for training and test samples

Table 3.11 Comparison of Prediction Errors Using Different Methods for The Jenkins-Box Data

Model name	Number of inputs	MSE
ANFIS model [100]	2	0.0073
FuNN model [101]	2	0.0051
FNT model (case 1) [190]	2	0.00066
FNT model (case 2) [190]	7	0.00029
HB-Spline network	5	0.0012

3.4   Hierarchical Wavelet Neural Networks

We illustrate an Hierarchical Wavelet Neural Network (HWNN) framework, which is highly suitable for function approximation problems. Based on the pre-defined instruction/operator sets, an HWNN network can be created and evolved, which also allows input variables selection. The proposed method interleaves both optimizations. Starting with random structures and corresponding parameters, it first tries to improve the structure and then as soon as an improved structure is found, it fine tunes its parameters. It then goes back to improving the structure again and, fine tunes the structure and parameters. This loop continues until a satisfactory solution is found or a time limit is reached.

3.4.1   Wavelet Neural Network

In terms of wavelet transformation theory, wavelets are usually described in the following form

$$\Psi = \{\Psi_i = |\mathbf{a_i}|^{-\frac{1}{2}}\psi(\frac{\mathbf{x} - \mathbf{b_i}}{\mathbf{a_i}}) : \mathbf{a_i}, \mathbf{b_i} \in R^n, i \in Z\}, \qquad (3.11)$$

where $\mathbf{x} = (x_1, x_2, \ldots, x_n)$, $\mathbf{a_i} = (a_{i1}, a_{i2}, \ldots, a_{in})$, $\mathbf{b_i} = (b_{i1}, b_{i2}, \ldots, b_{in})$ are a family of functions generated from one single function $\psi(x)$ by the operation of dilation and translation. $\psi(x)$, which is localized in both the time space and the frequency space, is called a mother wavelet and the parameters a_i and b_i are named the scale and translation parameters, respectively. The $\mathbf{x}$ represents inputs to the WNN model.

In the standard form of wavelet neural network, the output of a WNN is given by:

$$f(x) = \sum_{i=1}^{M} \omega_i \Psi_i(x) = \sum_{i=1}^{M} \omega_i |a_i|^{-\frac{1}{2}} \psi(\frac{x - b_i}{a_i}) \qquad (3.12)$$

where ψ_i is the wavelet activation function of ith unit of the hidden layer and ω_i is the weight connecting the ith unit of the hidden layer to the output layer unit. Note that for the n-dimensional input space, the multivariate wavelet basis function can be calculated by the tensor product of n single wavelet basis functions as $\psi(x) = \prod_{i=1}^{n} \psi(x_i)$.

3.5 Automatic Design of Hierarchical Wavelet Neural Network

In order to generate and optimize a mutilevel HWNN model, a tree-structural representation is adopted. For generating the tree, a function set F and a terminal instruction set T are described as $S = F \bigcup T = \{+_2, +_3, \ldots, +_N\} \bigcup \{x_1, \ldots, x_n\}$, where $+_i (i = 2, 3, \ldots, N)$ denote non-leaf nodes' instructions and taking i arguments. $x_1, x_2, \ldots, x_n$ are leaf nodes' instructions and taking no other arguments. The output of a non-leaf node is calculated as a wavelet neural network model by Eqn. (3.12). From this point of view, the instruction $+_i$ is also called a WNN operator with i inputs.

The WNN operator is depicted in Figure 3.11 (left). The mother wavelet $\psi(x) = -x exp(-\frac{x^2}{2})$ is used for experimental illustrations, and the number of wavelet basis functions in hidden layer is same with the number of inputs, that is, $m = n$.

In the construction process of HWNN tree, if a nonterminal instruction, i.e., $+_i (i = 2, 3, 4, \ldots, N)$ is selected, i real values are randomly generated and used for representing the connection strength between the node $+_i$ and its children. In addition, $2 \times n^2$ adjustable parameters a_i and b_i are randomly created as dilation and translation parameters of the wavelet basis functions. The output of the node $+_i$ can be calculated by using Eqn. (3.2). The overall output of HWNN tree can be computed from left to right by depth-first method, recursively.

3.5.1 Ant Programming for Evolving the Architecture of HWNN

Finding an optimal or near-optimal HWNN is formulated as a product of evolution [195]. For the structure optimization, any kind of tree structure based evolutionary algorithms can be employed. We illustrate Ant Programming (AP) [199] to find an optimal or near-optimal HWNN structure.

Ant programming is a new method, which applies the principle of the ant systems to automated program synthesis [199]. In the AP algorithm, each ant will build and modify the trees according to the quantity of pheromone at each node. The pheromone table appears as a tree. Each node owns a table which memorize the rate of pheromone to various possible instructions (Figure 3.12).

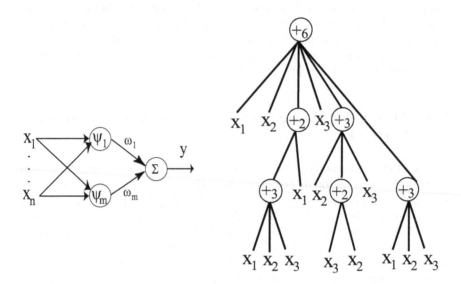

Fig. 3.11 A WNN operator (left), and a tree-structural representation of a HWNN with function instruction set $F = \{+_2, +_3, +_4, +_5, +_6\}$, and terminal instruction set $T = \{x_1, x_2, x_3\}$ (right)

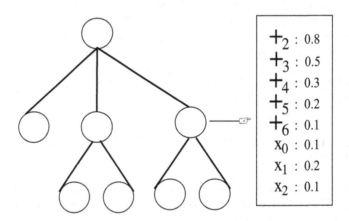

Fig. 3.12 Pheromone tree, in each node a pheromone table holds the quantity of pheromone associated with all possible instructions

First, a population of programs are generated randomly. The table of pheromone at each node is initialized at 0.5. This means that the probability of choosing each terminal and function is equal initially. The higher the rate of pheromone, the higher the probability to be chosen. Each program (individual) is then evaluated using a predefined objective function. The table of pheromone is update by two mechanisms:

(1) Evaporation decreases the rate of pheromone table for every instruction on every node according to following formula :

$$P_g = (1 - \alpha)P_{g-1} \qquad (3.13)$$

where P_g denotes the pheromone value at the generation g, α is a constant ($\alpha = 0.15$).

(2) For each tree, the components of the tree are reinforced according to the fitness of the tree. The formula is:

$$P_{i,s_i} = P_{i,s_i} + \frac{\alpha}{Fit(s)} \qquad (3.14)$$

where s is a solution (tree), $Fit(s)$ its fitness, s_i the function or the terminal set at node i in this individual, α is a constant ($\alpha = 0.1$), P_{i,s_i} is the value of the pheromone for the instruction s_i in the node i.

A brief description of AP algorithm is as follows:

S1 Every component of the pheromone tree is set to an average value;
S2 Random generation of tree based on the pheromone tree;
S3 Evaluation of ants using fitness function;
S4 Update of the pheromone table according to Eqn. 3.13 and Eqn. 3.14;
S5 Go to step S1 unless some criteria is satisfied.

3.5.2 Parameter Optimization Using Differential Evolution Algorithm

Parameter optimization is achieved using the differential evolution (DE) algorithm. DE works as follows: First, all individuals are randomly initialized and evaluated using the fitness function provided. Afterwards, the following process will be executed as long as the termination condition is not fulfilled: For each individual in the population, an offspring is created using the weighted difference of parent solutions. The offspring replaces the parent if it is fitter. Otherwise, the parent survives and is passed on to the next iteration of the algorithm.

3.5.3 Procedure of the General Learning Algorithm for HWNN

The general learning procedure for constructing the HWNN network can be described as follows:

1) Create an initial population randomly (HWNN trees and its correspond-
    ing parameters);
2) Structure optimization is achieved by using ant programming algorithm;
3) If a better structure is found, then go to step 4), otherwise go to step 2);
4) Parameter optimization is achieved by the DE algorithm as described in
    subsection 2. In this stage, the architecture of HWNN model is fixed, and
    it is the best tree developed during the end of run of the structure search.
    The parameters (weights and flexible activation function parameters) en-
    coded in the best tree formulate a particle.
5) If the maximum number of local search is reached, or no better parameter
    vector is found for a significantly long time then go to step 6); otherwise
    go to step 4);
6) If satisfactory solution is found, then the algorithm is stopped; otherwise
    go to step 2).

3.5.4 Variable Selection Using HWNN Paradigms

It is often a difficult task to select important variables for a classification or
regression problem, especially when the feature space is large. Conventional
HWNN usually cannot do this. In the perspective of HWNN framework, the
nature of model construction procedure allows the HWNN to identify important input features in building an HWNN model that is computationally
efficient and effective. The mechanisms of input selection in the HWNN constructing procedure are as follows. (1) Initially the input variables are selected
to formulate the HWNN model with same probabilities; (2) The variables
which have more contribution to the objective function will be enhanced and
have high opportunity to survive in the next generation by a evolutionary
procedure; (3) The evolutionary operators i.e., crossover and mutation, provide a input selection method by which the HWNN should select appropriate
variables automatically.

3.5.5 Experimental Illustrations

Application to Mackey-Glass Time-Series

The chaotic Mackey-Glass differential delay equation is recognized as a benchmark problem that has been used and reported by a number of researchers
for comparing the learning and generalization ability of different models. The
mackey-Glass chaotic time series is generated from the following equation:

$$\frac{dx(t)}{dt} = \frac{ax(t-\tau)}{1+x^{10}(t-\tau)} - bx(t). \tag{3.15}$$

Where $\tau > 17$, the equation shows chaotic behavior.

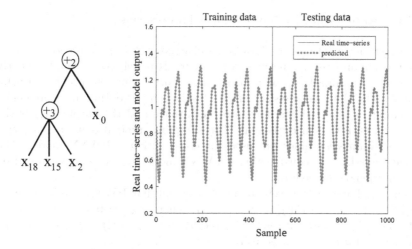

Fig. 3.13 The evolved architecture of HWNN model for prediction of the Mackey-Glass time-series(left), and the actual time series data, output of the evolved HWNN model and the prediction error(right)

Table 3.12 Comparison of prediction error using different methods for the Mackey-Glass time-series problem

Model name and reference	Number of inuts	RMSE
FNT model (Case 1) [190]	4	0.0069
FNT model (Case 2) [190]	7	0.0027
HWNN model	4	0.0043

To make the comparison with earlier work fair, we attempt to predict $x(t+6)$ by using the input variables $x(t)$, $x(t-1)$, ..., $x(t-18)$. 1000 sample points were used and the first 500 data pairs of the series were used as training data, while the remaining 500 were used to validate the model identified.

The used instruction sets to create an optimal HWNN model is $S = F \bigcup T = \{+_2, +_3, +_4\} \bigcup \{x_0, x_1, \ldots, x_{18}\}$, where $x_i (i = 0, 1, \ldots, 18)$ denotes $x(t)$, $x(t-1)$, ... $x(t-18)$.

After 12 generations of the evolution, an optimal HWNN model was obtained with RMSE 0.0045. The RMSE value for validation data set is 0.0043. The evolved HWNN is depicted in Figure 3.13 (left). The actual time-series data, the output of HWNN model are illustrated in Figure 3.13 (right). From the evolved HWNN tree, it is evident that the optimal inputs variables for constructing a HWNN model are: $x(t)$, $x(t-2)$, $x(t-15)$ and $x(t-18)$. A comparison result of different methods for forecasting Mackey-Glass data is illustrated in Table 3.12.

3.5.6 Application to Jenkins-Box Time-Series

The gas furnace data (series J) of Box and Jenkins (1970) was recorded from a combustion process of a methane-air mixture. It is well known and frequently used as a benchmark example for testing identification and prediction algorithms. The data set consists of 296 pairs of input-output measurements. The input $u(t)$ is the gas flow into the furnace and the output $y(t)$ is the CO_2 concentration in outlet gas. The sampling interval is 9s. For illustration, 10 inputs variables are used for constructing a HWNN model. The proper time-lags for constructing a HWNN model are finally determined by an evolutionary procedure.

The used instruction sets to create an optimal HWNN model is $S = F \bigcup T = \{+_2, \ldots, +_4\} \bigcup \{x_0, x_1, \ldots, x_9\}$, where $x_i(i = 0, 1, \ldots, 9)$ denotes $y(t-1)$, $y(t-2)$, $y(t-3)$, $y(t-4)$, and $u(t-1)$, $u(t-2)$, $u(t-3)$, $u(t-4)$, $u(t-5)$ and $u(t-6)$, respectively.

After 21 generations of the evolution, the optimal HWNN model was obtained with MSE 0.00021. The MSE value for validation data set is 0.00025. The evolved HWNN is depicted in Figure 3.14 (left) and the actual time-series, the HWNN model output and the prediction error is illustrated in Figure 3.14 (right). From the evolved HWNN tree, it is evident that the optimal inputs variables for constructing a HWNN model are: $u(t-3)$, $u(t-4)$, $u(t-6)$, $y(t-1)$, $y(t-2)$ and $y(t-3)$. It should be noted that the HWNN model with proper selected input variables has better precision and good generalization ability. A comparison result of different methods for forecasting Jenkins-Box data is illustrated in Table 3.13.

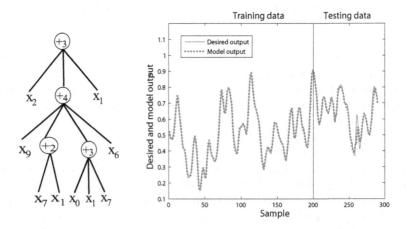

Fig. 3.14 The evolved structure of HWNN model for prediction of Jenkins-Box data (left), and the actual time series data, output of the evolved HWNN model for training and test samples (right)

Table 3.13 Comparison of prediction errors using different methods for the gas furnace data

Model name and reference	Number of inuts	MSE
FNT model (case 1) [190]	2	0.00066
FNT model (case 2) [190]	7	0.00029
HWNN model	6	0.00025

From the above simulation results, it is clear that the proposed HWNN model works well for function approximation problems.

Part IV

Hierarchical Fuzzy Systems

Hierarchical Fuzzy Systems

4.1 Introduction

Fuzzy inference systems [200], [201], [225] have been successfully applied to a number of scientific and engineering problems during recent years. The advantage of solving complex nonlinear problems by utilizing fuzzy logic methodologies is that the experience or expert's knowledge described as the fuzzy rule base can be directly embedded into the system for dealing with the problems. Many efforts have been made to enhance systematic design of fuzzy logic systems [203], [204], [205], [206], [207], [239], [244]. Some research focus on automatically finding the appropriate structure and parameters of fuzzy logic systems by using genetic algorithms [204], [207], [239], evolutionary programming [206], tabu search [208], and so on. There are many research works focusing on partitioning of the input space, to determine the fuzzy rules and parameters evolved in the fuzzy rules for a single fuzzy system [232], [229]. As it is well known, the curse-of-dimensionality is an unsolved problem in the fields of fuzzy and/or neuro-fuzzy systems [243].

Some of the problems mentioned above are partially solved by several researchers working in the hierarchical fuzzy systems domain [208], [209], [210], [211], [212], [213], [214], [215], [216], [227]. Torra [202] has summarized all the related recent research. As a way to overcome the curse-of-dimensionality, it was suggested by Brown et al. [216] to arrange several low-dimensional rule base in a hierarchical structure, i.e., a tree, causing the number of possible rules to grow in a linear way according to the number of inputs. A method was proposed to determine automatically the fuzzy rules in a hierarchical fuzzy model [231]. Rainer [214] described a new algorithm which derives the rules for hierarchical fuzzy associative memories that were structured as a binary tree. Wang and Wei [210], [211], [217] proposed specific hierarchical fuzzy systems and its universal approximation property was proved. But the main problem lies in fact that this is a specific hierarchical fuzzy system which lacks flexibility in structure adaptation, and it is difficult to arrange the input variables for each sub-model. Lin and Lee [218] proposed a genetic

Y. Chen, A. Abraham.: Tree-Struc. Based Hybrid Com. Intelligence, ISRL 2, pp. 129–147.
springerlink.com

algorithm based approach to optimize the hierarchical structure and the parameters of 5-inputs hierarchical fuzzy controller for the low-speed control problem. Based on the analysis of importance of each input variable and the coupling between any two input variables, the problem of how to distribute the input variables to different (levels of) relational modules for incremental and aggregated hierarchical fuzzy relational systems was addressed [227].

Building a hierarchical fuzzy system is a difficult task. This is because we need to define the architecture of the system (the modules, the input variables of each module, and the interactions between modules), as well as the rules of each modules. Two approaches could be used to tackle this problem. One approach is that an expert supplies all the required knowledge for building the system. The other one is to use machine and/or optimization techniques to construct/adapt the system. Several machine learning and optimization techniques have been applied to aid the process of building hierarchical fuzzy systems. For example, Shimojima et al. use genetic algorithm to determine the hierarchical structure [231]. This is combined with backpropagation and gradient descent algorithm to fine tune its parameters. A structure identification method of sub-models for hierarchical fuzzy modeling using the multiple objective genetic algorithm was proposed by Tachibana and Furuhashi [233]. Chen et al. [234] have proposed a hybrid method using ant programming and particle swarm optimization algorithm to optimize the hierarchical TS fuzzy model.

From now onwards, the hierarchical structure or structure for short means the way of arrangement of hierarchical TS fuzzy systems and the position/selection of each input variable in the sub-fuzzy systems. The free parameters to be optimized including all the parameters used in the hierarchical TS fuzzy systems including membership function parameters for each fuzzy sets, the free parameters in the consequent parts of the fuzzy rule base for each sub-fuzzy systems.

We illustrate a systematic design method for the hierarchical TS-FS model. The hierarchical structure is evolved using a Probabilistic Incremental Program Evolution (PIPE) [219], [220], [221] with specific instructions, an algorithm originally used for automatic program synthesis. The fine tuning of the rule's parameters encoded in the structure is accomplished using Evolutionary Programming (EP). The proposed method interleaves both PIPE and EP optimizations. Starting with random structures and rules' parameters, it first tries to improve the hierarchical structure and then as soon as an improved structure is found, it fine tunes its rules' parameters. It then goes back to improve the structure again and, provided it finds a better structure, it again fine tunes the rules' parameters. This loop continues until a satisfactory solution (hierarchical TS-FS model) is found or a time limit is reached. The novelty of this paper is in the usage of evolutionary mechanism for selecting the important features and for constructing a hierarchical TS fuzzy model automatically.

The rest of the Chapter is organized as follows. A new encoding and an automatic design method for the hierarchical TS-FS is illustrated with some simulation results related to system identification and time-series prediction problems.

4.2   Takagi-Sugeno Fuzzy Inference System (TS-FS)

Fuzzy inference systems are composed of a set of *if-then* rules. A Takagi-Sugeno fuzzy model has the following form of fuzzy rules [1] :

$$R_j : \text{if } x_1 \text{ is } A_{1j} \text{ and } x_2 \text{ is } A_{2j} \text{ and } \dots \text{ and } x_n \text{ is } A_{nj}$$
$$\text{Then } y = g_j(x_1, x_2, \dots, x_n), (j = 1, 2, \dots, N)$$

where $g_j(\cdot)$ is a crisp function of x_i. Usually, $g_j(x_1, x_2, \dots, x_n) = \omega_0 + \omega_1 x_1 + \omega_2 x_2 + \cdots + \omega_n x_n$. The overall output of the fuzzy model is obtained by:

$$y = \frac{\sum_{j=1}^{N} g_j(\cdot) T_{i=1}^{m_j} \mu_{ij}(x_i)}{\sum_{j=1}^{N} T_{i=1}^{m_j} \mu_{ij}(x_i)} \tag{4.1}$$

where $1 \leq m_j \leq n$ is the number of input variables that appear in the rule premise, N is the number of fuzzy rules, n is the number of inputs, μ_{ij} is the membership function for fuzzy set, A_{ij} and T is a T-norm for fuzzy conjunction.

The TS-FS is a single-stage fuzzy system. It is important to partition the input space using some clustering, grid partitioning etc. [228]. The shapes of membership functions in the antecedent parts, and the free parameters in the consequent parts are also to be determined using some adaptive techniques [229], [230], [244].

4.3   Hierarchical TS-FS: Encoding and Evaluation

An hierarchical fuzzy inference system not only provides a more complex and flexible architecture for modelling nonlinear systems, but can also reduce the size of the rule base to some extend. Figure 4.1 depicts some possible hierarchical TS-FS models for 4 input variables and 3 hierarchical layers. The problems in designing a hierarchical fuzzy logic system includes the following:

- Selecting an appropriate hierarchical structure;
- Selecting the inputs for each fuzzy TS sub-model;
- Determining the rule base for each fuzzy TS sub-model;
- Optimizing the parameters in the antecedent parts and the linear weights in the consequent parts.

There is no direct/systematic method for designing the hierarchical TS-FS. From the evolution point of view, finding a proper hierarchical TS-FS model can be posed as a search problem in the structure and parameter space. For this purpose, a new encoding method for hierarchical TS-FS is developed.

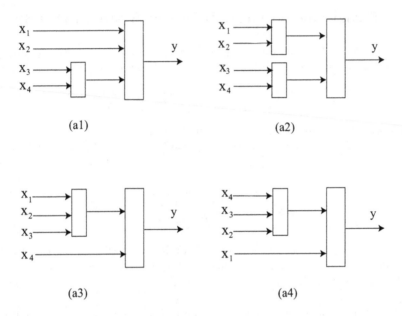

Fig. 4.1 Example of possible hierarchical fuzzy logic models, number of inputs: 4, number of layers: 3

4.3.1   Encoding

A tree-structural based encoding method with specific instruction set is selected for representing a hierarchical TS-FS in this research. The reasons for choosing this representation are that (1) the tree has a natural and typical hierarchical layer; (2) with pre-defined instruction sets, the tree can be created and evolved using the existing tree-structure-based approaches, i.e., Genetic Programming (GP) and PIPE algorithms.

Assume that the used instruction set is $I = \{+_2, +_3, x_1, x_2, x_3, x_4\}$, where $+_2$ and $+_3$ denote non-leaf nodes' instructions taking 2 and 3 arguments, respectively. x_1, x_2, x_3, x_4 are leaf nodes' instructions taking zero arguments each. In addition, the output of each non-leaf node is calculated as a single TS fuzzy sub-model. For this reason the non-leaf node $+_2$ is also called a two-input TS fuzzy instruction/operator. Figure 2 illustrates the tree structural representation of the hierarchical TS fuzzy models (as per in Figure 1.)

It should be noted that in order to calculate the output of each TS fuzzy sub-model (non-leaf node), parameters in the antecedent parts and consequent parts of the TS fuzzy sub-model should be embedded into the tree.

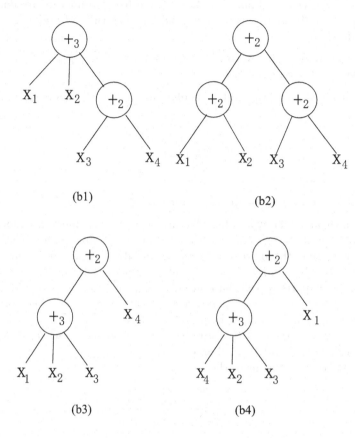

(b1)

(b2)

(b3)

(b4)

Fig. 4.2 Tree structural representation of the hierarchical T-S fuzzy models as shown in Figure 1 (a1), (a2), (a3) and (a4), where the used instruction set is $I = \{+_2, +_3, x_1, x_2, x_3, x_4\}$.

4.3.2 Evaluation

The output of a hierarchical TS-FS tree can be calculated from a layer to layer basis. For simplicity, the calculation process of the tree (Figure 4.2 (b1)) is illustrated below. Assume that each input variable is divided into two fuzzy sets and the used fuzzy membership function is:

$$\mu(a, b; x) = \frac{1}{1 + (\frac{x-a}{b})^2} \tag{4.2}$$

First, the output of the TS fuzzy sub-model (node +2) is computed. Assume that the used fuzzy sets for variables $x3$ and $x4$ are A_{11}, A_{12} and A_{21}, A_{22}, respectively. Suppose that the parameters in the consequent parts of rule base are c_{ij}^0, c_{ij}^1, c_{ij}^2, ($i = 1, 2$ and $j = 1, 2$). These free parameters are encoded in the node +2. Therefore, the corresponding fuzzy rules of node +2 can be described as:

$R_{i,j}$: if $x3$ is A_{1i} and $x4$ is A_{2j} then $y_{ij} = c_{ij}^0 + c_{ij}^1 x3 + c_{ij}^2 x4$
for $i = 1, 2$ and $j = 1, 2$.

The output of node +2 can be calculated based on the TS fuzzy model:

$$y = \frac{\sum_{i=1}^2 \sum_{j=1}^2 \sigma_{ij} y_{ij}}{\sum_{i=1}^2 \sum_{j=1}^2 \sigma_{ij}} \tag{4.3}$$

where

$$\sigma_{ij} = \mu_{A_{1i}}(x3)\mu_{A_{2j}}(x4) \text{ for } i = 1, 2 \text{ and } j = 1, 2.$$

Second, the overall output of the hierarchical TS fuzzy model is computed. It has three inputs, x_1, x_2 and y, the output of the TS fuzzy sub-model (node +2). Assume that the used fuzzy sets for variables x_1, x_2 and y are: B_{11}, B_{12}, B_{21}, B_{22}, B_{31} and B_{32}, respectively. Suppose that the parameters in the consequent parts of rule base are d_{ijl}^0, d_{ijl}^1, d_{ijl}^2, and d_{ijl}^3 ($i = 1, 2, j = 1, 2, l = 1, 2$). These free parameters are encoded in node +3. The complete fuzzy rules of node +3 can be described as follows:

$R_{i,j,l}$: if x_1 is B_{1i}, x_2 is B_{2j}, y is B_{3l} then $z_{ijl} = d_{ijl}^0 + d_{ijl}^1 x_1 + d_{ijl}^2 x_2 + d_{ijl}^3 y$ for $i = 1, 2$, and $j = 1, 2$ and $l = 1, 2$.
Thus, the overall output of the tree is

$$z = \frac{\sum_{i=1}^2 \sum_{j=1}^2 \sum_{l=1}^2 \mu_{ijl} z_{ijl}}{\sum_{i=1}^2 \sum_{j=1}^2 \sum_{l=1}^2 \mu_{ijl}} \tag{4.4}$$

where

$$\mu_{ijl}(x_1, x_2, y) = \mu_{B_{1i}}(x_1)\mu_{B_{2j}}(x_2)\mu_{B_{3l}}(y).$$

It should be noted that all the parameters encoded in the tree are randomly generated along with the creation of the tree initially, which will be further optimized using evolutionary programming.

4.3.3 Objective Function

The fitness function used for the PIPE and EP is given by mean square error (MSE):

$$Fit(i) = \frac{1}{P} \sum_{j=1}^P (y_1^j - y_2^j)^2 \tag{4.5}$$

or Root Mean Square Error ($RMSE$):

$$Fit(i) = \sqrt{\frac{1}{P}\sum_{j=1}^{P}(y_1^j - y_2^j)^2} \tag{4.6}$$

where P is the total number of samples, y_1^j and y_2^j are the actual and hierarchical TS-FS model outputs of j-th sample. $Fit(i)$ denotes the fitness value of the i-th individual.

4.4 Evolutionary Design of Hierarchical TS-FS

The hierarchical structure is created and optimized using PIPE with specific instructions and the fine turning of the rule's parameters encoded in the structure is accomplished using EP algorithm.

4.4.1 Algorithm for Designing Hierarchical TS-FS Model

Combining the self-organizing and structure learning characteristics of PIPE and the parameter optimization ability of EP, we propose the following hybrid algorithm for designing the hierarchical TS-FS model (Figure 4.3).

1) Set the initial values of parameters used in the PIPE and EP algorithms. Set the elitist program as NULL and its fitness value as a biggest positive real number of the computer at hand. Create the initial population (tree) and corresponding parameters used in hierarchical TS-FS model.

2) Do structure optimization using PIPE algorithm, in which the fitness function is calculated by Eqn. (4.5) or Eqn. (4.6).

3) If a better structure is found, then go to step 4), otherwise go to step 2). The criterion concerning with better structure found is distinguished as follows: if the fitness value of the best program is smaller than the fitness value of the elitist program, or the fitness values of two programs are equal but the nodes of the former is lower than the later, then we say that the better structure is found.

4) Parameter optimization using EP search. In this step, the tree structure or architecture of hierarchical TS-FS model is fixed, and it is the best tree taken from the end of run of PIPE search. All of the rules' parameters encoded in the best tree will be optimized by EP search in order to decrease the fitness value of best program.

5) If the maximum number of EP search is reached, or no better parameter vector is found for a significantly long time (100 steps) then go to step 6); otherwise go to step 4).

6) If satisfactory solution is found, then stop; otherwise go to step 2).

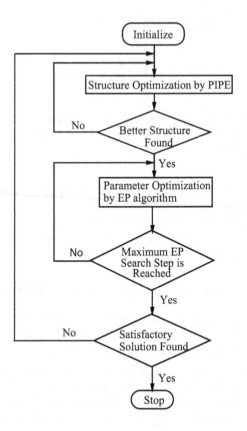

Fig. 4.3 The flow chart of the proposed algorithm for designing the hierarchical TS-FS model

4.4.2 Feature/Input Selection with Hierarchical TS-FS

It is often a difficult task to select important variables for prediction and classification problems, especially when the feature space is large. A predefined single/multi level fuzzy model usually cannot do this. In the perspective of evolution-driven hierarchical TS-FS framework, the nature of model construction procedure allows the H-TS-FS to identify important input features in building an prediction or classification model that is computationally efficient and effective. The mechanisms of input selection in the H-TS-FS constructing procedure are as follows.

- Initially the input variables are selected to formulate the Hierarchical TS-FS model with same probabilities;

- The variables that has more contribution to the objective function will be enhanced and have high opportunity to survive in the next generation by a evolutionary procedure;
- The evolutionary operators i.e., crossover and mutation, provide a input selection method by which the Hierarchical TS-FS would select the appropriate variables automatically.

4.5 Experimental Illustrations

The proposed approach has been evaluated for nonlinear system identification problems, Mackey-Glass chaotic time-series prediction problem, and the Iris and Wine classification problems. The next subsections discuss these applications and the results obtained by the evolutionary design of hierarchical TS-FS model and the performance is compared with other fuzzy/neural learning approaches.

Table 4.1 Parameters Used in the PIPE Algorithm

Parameters	Values
population size PS	100
elitist learning probability P_{el}	0.01
learning rate lr	0.01
fitness constant ε	0.000001
overall mutation probability P_M	0.4
mutation rate mr	0.4

The used parameters in PIPE are shown in Table 4.1. The parameters used in EP: population size is 60, opponent number $Q = 30$, $\alpha = 0.3$. For all the simulations, the minimum and maximum number of hierarchical layers are predefined as 2 and 4 and each input variable is partitioned into 2 fuzzy sets. The used fuzzy membership function is shown in Eqn.(4.2). The initial fuzzy rules for each sub-fuzzy systems are randomly generated and all the free parameters including fuzzy sets membership function parameters and the free parameters in the consequent parts of fuzzy rules are randomly generated at [0,1] initially. It should be noted that the selection of the non-leaf's instruction is experimental. Selecting more instructions will increase the structure/parameter search space and results in a bigger hierarchical TS fuzzy system. For an identification or classification problem, if the input number is n, selecting the maximum instruction $+_N$ as $N = n/3$ is enough according to our experiments. This experimental rule should reduce the search space significantly.

4.5.1   Systems Identification

The first plant to be identified is a linear system given by [224],

$$y(k+1) = 2.627771y(k) - 2.333261y(k-1) + 0.697676y(k-2)$$

$$+0.017203u(k) - 0.030862u(k-1) + 0.014086u(k-2) \quad (4.7)$$

400 data points were generated with the randomly selected input signal $u(k)$ between -1.0 and 1.0. The first 200 points were used as training data set and the remaining data were used as validation data set. The input vector is set as $x = [y(k), y(k-1), y(k-2), u(k), u(k-1), u(k-2)]$. The used instruction set is $I = \{+_2, +_3, +_4, x_0, x_1, x_2, x_3, x_4, x_5\}$.

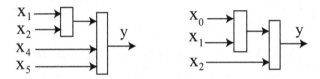

Fig. 4.4 The structure of evolved hierarchical TS-FS models, (left) for plant 1, and (right) for plant 2

10 independent runs were taken. The average training time for 10 runs is 245 seconds. The best structure of evolved hierarchical TS-FS models is shown in Figure 4.4(left). The output of the evolved model, the actual output and the test error for test data set are illustrated in Figure 4.5.

The second plant to be identified is a nonlinear system given by [224]:

$$y(k+1) = \frac{y(k)}{1.5 + y^2(k)} - 0.3y(k-1) + 0.5u(k) \quad (4.8)$$

The input and output of system are $x(k) = [u(k), u(k-1), y(k), y(k-1)]$ and $y(k+1)$, respectively.

The training samples and the test data set are generated by using the same sequence of random input signals as mentioned previously. The used instruction set is $I = \{+_2, +_3, +_4, x_0, x_1, x_2, x_3\}$.

10 independent runs were run. The average training time of 10 runs is 317 seconds. The best structure of evolved hierarchical TS-FS model is shown in Figure 4.4(right). The output of the evolved model, the actual output and the test error for test data set are shown in Figure 4.6.

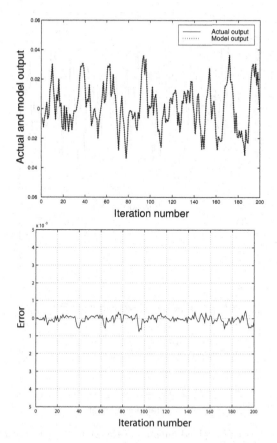

Fig. 4.5 The actual and evolved outputs of the plant 1 for the test data set (left), and the test error (right)

For comparison, the test results obtained by Elman and Jordan neural networks [224], Wavelet Neural Networks (WNN) [190] and the proposed Hierarchical TS-FS model (H-TS-FS) are shown in Table 4.2. From the above simulation results, it is evident that the proposed hierarchical TS-FS model works very well for identifying the linear/nonlinear systems much better than the neural network models.

4.5.2 Chaotic Time-Series of Mackey-Glass

The Mackey-Glass chaotic differential delay equation is recognized as a benchmark problem that has been used and reported by a number of researchers for comparing the learning and generalization ability of different models. The Mackey-Glass chaotic time series is generated using the following differential equation:

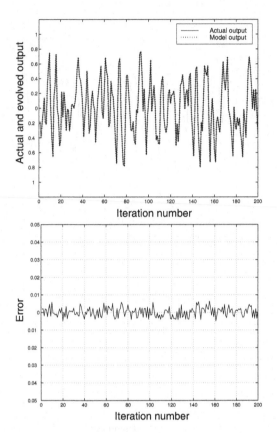

Fig. 4.6 The actual and evolved outputs of the plant 2 for the test data set (left), and the test error (right)

Table 4.2 The comparison of the MSE values for modified Elman nets [224], modified Jordan nets [224], wavelet neural networks (WNN)[190] and hierarchical TS-FS model for test data set

Plant	Elman	Jordan	WNN	H-TS-FS
1	0.0000548	0.0000492	0.000000459	0.0000000432
2	0.0004936	0.0003812	0.000002728	0.0000007065

$$\frac{dx(t)}{dt} = \frac{ax(t-\tau)}{1+x^{10}(t-\tau)} - bx(t). \tag{4.9}$$

where $a = 0.2$ and $b = 0.1$, $\tau > 17$ the equation shows chaotic behavior. In our simulations, $\tau = 30$ has been adopted. To compare with previous works [227], we predicted the value of $x(t+6)$ using the input variables $x(t-30)$,

$x(t-24)$, $x(t-18)$, $x(t-12)$, $x(t-6)$ and $x(t)$, where $t=130$ to $t=1129$. It corresponds to a 6-input to 1-output mapping.

1000 sample points were used in our study. The first 500 data pairs were used as training data, while the remaining 500 were used to validate the model identified. The used instruction set is $I = \{+_2, +_3, \ x_0, x_1, x_2, x_3, x_4, x_5\}$, where $x_0, x_1, x_2, x_3, x_4, x_5$ denote $x(t-30)$, $x(t-24)$, $x(t-18)$, $x(t-12)$, $x(t-6)$ and $x(t)$, respectively.

The results are obtained from training the hierarchical TS-FS models using 10 different experiments. The average training time of 10 runs is 719 seconds. The average $RMSE$ value for training and test data sets are 0.017 and 0.015, respectively.

Two evolved structures of hierarchical TS-FS models are shown in Figure 4.7. A comparison has been made to illustate the actual time-series, the hierarchical TS-FS model output and the prediction error (Figure 4.8). Figure 4.8 also depicts the convergence performance of the best hierarchical TS-FS model. Performance comparison of the different methods for approximating the Mackey-Glass data is shown in Table 4.3.

Table 4.3 Comparison of the incremental type multilevel FRS (IFRS) [227], the aggregated type mutilevel FRS (AFRS) [227], and the hierarchical TS-FS in Mackey-Glass time-series prediction (H-TS-FS1 and H-TS-FS2 are corresponding to the model structures shown in Figure - right and left, respectively)

Model	Stage	No. of rules	No. of parameters	RMSE(train)	RMSE(Test)
IFRS	4	25	58	0.0240	0.0253
AFRS	5	36	78	0.0267	0.0256
H-TS-FS1	3	28	148	0.0120	0.0129
H-TS-FS2	2	12	46	0.0145	0.0151

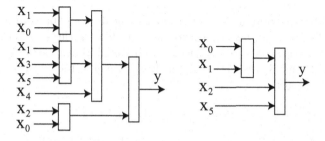

Fig. 4.7 Two possible structures of hierarchical TS-FS models for predicting the Mackey-Glass time-series: with RMSE = 0.01205 (left) and with RMSE = 0.01417 (right)

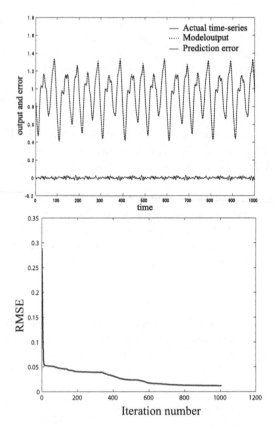

Fig. 4.8 Actual time-series, model output and prediction error for test data set (left), and fitness curve for training (right)

4.5.3   Iris Data Classification

The Iris data is a common benchmark in classification and pattern recognition research [235]. It contains 50 measurements of four features from each of the three species Iris setosa, Iris versicolor, and Iris virginica [237]. We label the species 1, 2, and 3, respectively, which gives a 5×150 pattern matrix of observation vectors:

$$Z_k^T = [x_1^k, x_2^k, x_3^k, x_4^k, c_k], c_k \in 1, 2, 3, k = 1, 2, \ldots, 150 \qquad (4.10)$$

where $x_1^k, x_2^k, x_3^k, x_4^k$ are the sepal length, sepal width, petal length, and petal width, respectively.

We normalized each attribute value into a real number in the unit interval. Table 4.4 shows the results of some well-known classifier systems. For the Iris example, we also used 150 patterns to design a hierarchical

Table 4.4 Comparison of results for Iris data

	Term sets	Rules	Recognition rate on total data set (%)
Wang et al. [238]	11	3	97.5
Wu et al. [239]	9	3	96.2
Shi et al. [204]	12	4	98.0
Russo [240]	18	5	100
Ishibuchi et al. [236]	7	5	98.0
HS-TS	-	16	99.6

Table 4.5 Results of ten runs on Iris data

	1	2	3	4	5	6	7	8	9	10	Average
Misclassification	1	1	0	0	0	1	1	1	1	0	0.6
Recognition rate (%)	99.3	99.3	100	100	100	99.3	99.3	99.3	99.3	100	99.6
Features	4	4	3	4	4	3	4	4	3	4	3.7
Rules	12	12	16	20	20	12	12	16	20	20	16
Parameters	64	60	84	108	108	60	64	84	104	108	84.4
Training time (minutes)	8	12	21	17	22	9	19	21	25	11	16.7

TS-FS classifier system via the proposed algorithm. The used instruction set is $F = \{+_2, +_3, x_1, x_2, x_3, x_4\}$.

Table 4.5 shows the results of ten runs (i.e. ten different initializations of parameters). To estimate the performance of the proposed method on unseen data, the five-fold cross-validation was performed on the iris data. In the five-fold cross-validation experiment, the normalized iris data were divided into five disjoint groups containing 30 different patterns each, with ten patterns belonging to each class. Then we derived the hierarchical TS-FS models via the proposed method on all data outside one group and tested the resulting hierarchical TS-FS classifier on the data within that group. Finally, five hierarchical TS-FS were derived. The evolved hierarchical architectures for five-fold cross-validation are shown in Figure 4.9. The convergence performance of five-fold cross validation test 3 is shown in Figure 11(left). Table 4.6 reports the results

Table 4.6 Five-Fold cross validation for Iris data

	1	2	3	4	5	Average (%)
Rules	12	20	20	24	12	17.6
Training patterns	120	120	120	120	120	120
Misclassification (training)	0	0	0	0	0	0
Recognition rate (training)(%)	100	100	100	100	100	100
Testing patterns	30	30	30	30	30	30
Misclassification (testing)	0	0	0	0	1	0.2
Recognition rate (testing)(%)	100	100	100	100	96.7	99.34

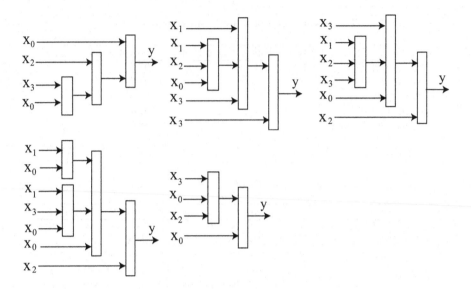

Fig. 4.9 The evolved optimal H-TS-FS architectures for five-fold cross-validation (Iris data)

Table 4.7 Comparison of results using Wine data

	Features	Term sets	Rules	Recognition rate on total data set (%)
Setnes et al. [241]	9	21	3	98.3
Wang et al. [238]	13	34	3	99.4
Roubos et al. [242]	5	15,11,10	3	98.9, 98.3. 99.4
Ishibuchi et al. [236]	-	9	6	100
HS-TS	4.9	-	16.4	99.6

of five-fold cross validation. The average classification result is 100.0% correct (no misclassifications) on the training data and 99.34% correct (average about 0.2 misclassification) on the test data using 17.6 rules (average).

4.5.4 Wine Data Classification

The wine data set is a 13-dimensional problem with 178 samples from three classes. We chose this data set because it involves many continuous attributes. We normalized each attribute value into a real number in the unit interval.

Table 4.7 illustrates the results of some well-known classifier systems.

For the wine data, we also used 178 patterns to design a hierarchical TS-FS classifier system using the HS-TS algorithm. The used instruction set is $F = \{+_2, +_3, +_4, x_1, x_2, \ldots, x_{13}\}$.

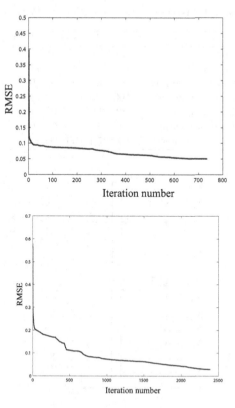

Fig. 4.10 The convergence performance of five-fold cross validation test 3 for Iris data (left), and for Wine data (right)

Table 4.8 Results of ten runs on Wine data

	1	2	3	4	5	6	7	8	9	10	Average
Misclassification	0	1	1	1	1	0	1	1	0	1	0.7
Recognition rate (%)	100	99.4	99.4	99.4	99.4	100	99.4	99.4	100	99.4	99.6
Features	5	4	4	5	5	6	4	6	6	4	4.9
Rules	16	12	12	20	16	20	12	20	20	16	16.4
Parameters	84	60	64	108	84	108	64	108	108	84	87.2
Training time (minutes)	10	14	23	19	24	13	22	24	28	18	19.5

Table 4.8 illustrates the empirical results of ten runs (i.e. ten different initializations of parameters). To estimate the performance of the proposed method on unseen data, the five-fold cross-validation was performed on the Wine data. In the five-fold cross-validation experiment, the normalized Wine data were divided into five disjoint groups. Then we derived the hierarchical TS-FS models using the proposed method on all data outside one group and

Table 4.9 Five-Fold cross validation for Wine data

	1	2	3	4	5	Average (%)
Rules	20	16	24	20	32	22.4
Training patterns	136	144	144	144	144	142.4
Misclassification (training)	0	0	0	0	0	0
Recognition rate (training)(%)	100	100	100	100	100	100
Testing patterns	42	34	34	34	34	35.6
Misclassification (testing)	0	1	0	0	0	0.2
Recognition rate (testing)(%)	100	97.1	100	100	100	99.4

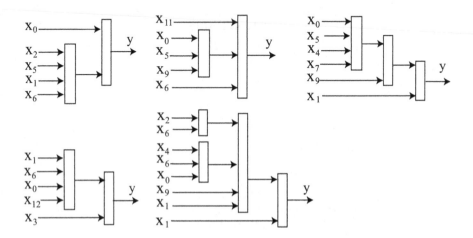

Fig. 4.11 The evolved optimal H-TS-FS architectures for five-fold cross-validation (Wine data)

tested the resulting hierarchical TS-FS classifier for the data within that group.

Finally, five hierarchical TS-FS were derived. The evolved hierarchical architectures for five-fold cross-validation are shown in Figure 4.11. The convergence performance of five-fold cross validation test is illustrated in Figure 4.10(right). Table 4.9 reports the results of five-fold cross validation. The average classification result is 100.0% correct (no misclassifications) on the training data and 99.4% correct (average about 0.2 misclassification) on the test data using 22.4 rules (average).

One major advantage of using a hierarchical TS-FS or a mutilevel fuzzy system other than a single-level system (direct approach) is that the number of fuzzy rules and fuzzy operations involved in modeling process can be reduced significantly when compared with those required by the single-level counterparts. Due to the limitations to solve the hierarchical TS-FS analytically, we choose to identify the hierarchical TS-FS using an evolutionary optimization approach.

First, the hierarchical structure and the rules' parameters can be flexibly encoded into a TS-FS tree. And then, the PIPE and the EP algorithms are employed to evolve the optimal structure and parameters alternatively. Other tree-structure based evolutionary algorithms and parameter learning algorithms can also be used to solve the problem. The methods used by IFRS and AFRS [227], the hierarchical structure and input selection are assigned based on: (1) analysis of the importance of each input variables; (2)analysis of the coupling between input variables. In contrast to the IFRS and AFRS, the hierarchical structure and input selection in this research are accomplished using an evolutionary procedure automatically. Furthermore, compared to the IFRS and AFRS the generated hierarchical TS-FS model has some advantages in terms of the approximation accuracy, the number of rules and the number of free parameters.

Simulation results shown that the evolved hierarchical TS-FS models are effective for the identification of linear/nonlinear systems, for the prediction of chaotic time-series, and for the classification of Iris and Wine data.

It should be noted that the hierarchical TS-FS has smaller number of rules than a single level (direct approach) TS-FS. The number of rules and parameters would increase tremendously (even difficult to manage) for large number of inputs if a direct approach is used. This also results in slow convergent speed.

Part V

Reverse Engineering of Dynamical Systems

5

Reverse Engineering of Dynamic Systems

5.1 Introduction

The general task of system identification problem is to approximate automatically the behavior of an unknown plant using an appropriate model. Identification of nonlinear system suffers many problems including determination of the structure and parameters of the system. Many traditional methods of system identification are based on parameter estimation, and mainly rely on least mean-squares (LMS) method. Recently soft computing based system identification approaches, i.e., neural networks and fuzzy systems have been an active research area.

For the identification and control problem of nonlinear dynamic system, the evolutionary identification/control has received much attention during the last few years [245][246][247][248][249][250][251]. These research works can be classified into two types of methods: the one is to identify the parameters of the system and the other is to identify the structure or whole model of the system. The former is illustrated by the Genetic Algorithms (GA), Evolutionary Programming (EP) and the latter by the Genetic Programming (GP) approach [245]. Kristinsson et al. provided an excellent account of the art in the area of parameter identification by using GA [251]. Andrew et.al. proposed a system identification method by using GP [245]. One of the unsolved problems of the evolutionary identification is that the structure and parameter of the system cannot be identified simultaneously. This Chapter is intended to the Tree-Structure-based Evolutionary Algorithm (TSEA) for system identification problem.

Tree-structure-based evolutionary algorithm is an ideal candidate for system identification and controller design, by the direct matching of individual structure and system model or control rule. The advantage of the method is that the evolved model is of symbolic expression and is easier to analyze than a black box model (e.g., a neural network model). This is a kind of structural evolution in general. But it is difficult to cope with the dynamic behavior of evolved system due to:

Y. Chen, A. Abraham.: Tree-Struc. Based Hybrid Com. Intelligence, ISRL 2, pp. 151–182.
springerlink.com

- the evolved symbolic expression is usually redundant and very long in length;
- there are some differences between the evolved symbolic expression and the traditional system representations, such as, transfer function, state space representation and differential equation.

Can we evolve a symbolic expression that can be represented as a meaningful expression, i.e., a differential equation or a transfer function and it can be easily addressed by using traditional techniques? Many studies have shown that the effectiveness of evolutionary algorithms depends sensitively on the choice of representations, that is, it is important for the problem at hand to choice a proper map between the search space of knowledge structures (the phenotype space) and the space of the chromosomes (the genotype space).

We illustrate a new representation scheme of the additive models, by which the linear and nonlinear system identification problems are addressed by using automatic evolutionary design procedure. The architecture of nonlinear systems and the input variables are evolved and selected by a tree-structure based evolutionary algorithm and the corresponding parameters are optimized by a random search algorithm.

The Chapter is organized as follows. First a gentle introduction to tree structural representation and calculation of the additive tree models is provided. Further an hybrid algorithm for evolving the additive tree models and some simulation results for the prediction of chaotic time series, the reconstruction of polynomials and the identification of the linear/nonlinear system is illustrated.

5.2   Calculation and Representation of Additive Models

Many scientific and engineering problems can be finally formulated as a complex nonlinear mapping problem, in which local linear models and local nonlinear models play a key role for analysis of the characteristics of the system. In order to efficiently analyze the characteristics of nonlinear systems, a method for evolutionary design of additive models is developed.

The candidate solution for analyzing the nonlinear system is represented as an additive tree model (Figure. 5.1), in which the root node returns the weighted sum of a number of linear/nonlinear terms according to the structure of the subtrees. The structure and weights/parameters of the additive tree are evolved by tree-structure based evolutionary algorithm and a random search algorithm, respectively.

Two instruction/operator sets I_0 and I_1 are used for generating the additive tree in this approach.

$$I_0 = \{+_2, +_3, \ldots, +_N\}$$
$$I_1 = F \cup T = \{*, \%, sin, cos, exp, rlog, x, R\}$$

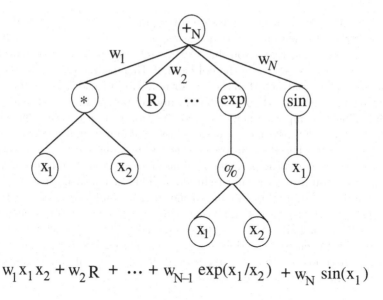

$$w_1 x_1 x_2 + w_2 R + \cdots + w_{N\text{-}1} \exp(x_1/x_2) + w_N \sin(x_1)$$

Fig. 5.1 Example: Interpretation of an additive tree and its symbolic expression

where $F = \{*, \%, sin, cos, exp, rlog\}$ and $T = \{x, R\}$ are function set and terminal set. $+_N$, $*$, $\%$, sin, con, exp, $rlog$, x, and R denote addition, multiplication, protected division ($\forall x, y \in R, y \neq 0 : x\%y = x/y$ and $x\%0 = 1$), sine, cosine, exponent, protected logarithm ($\forall x \in R, x \neq 0 : rlog(x) = log(abs(x))$ and $rlog(0) = 0$), system inputs, and random constant number, and taking N, 2, 2, 1, 1, 1, 1, 0 and -1 arguments respectively.

N is a integer number (the maximum number of linear/nonlinear terms), I_0 is the instruction set of the root node, and the instructions of other nodes are selected from instruction set I_1. Each node of $N_{d,w}$ of the tree contains a random constant $R_{d,w}$ or an instruction $I_{d,w}$, where d and w denote the node's depth and horizontal position of the tree.

Note that if the instruction set I_1 is defined as $I_1 = \{*_2, *_3, \ldots, *_n, x_1, x_2, \ldots, x_n, R\}$, the additive model becomes a polynomial in which the components and input variables of the polynomial can be selected and determined by using an evolutionary procedure.

5.3 Hybrid Algorithm

5.3.1 Tree-Structure Based Evolutionary Algorithm

Tree-structure based evolutionary algorithms include Genetic Programming (GP) [252] and Probabilistic Incremental Program Evolution (PIPE)[253][254] [255][256][257][258].

It is well known that an expression corresponding to $f_1(x) = x + x^2 + x^3$ is usually evolved much fast than one corresponding to $f_2(x) = 2.71x + 3.14x^2$ by GP for function regression due to the fact that it is difficult for GP to find the proper parameter in the polynomial. But if some real-valued parameters are added or attached to the tree or node and provide some appropriate calculation structure, GP can be fast in finding polynomials with real coefficients. This fact implies a new research direction to improve the efficiency of GP and to develop new tree-structure based computational models.

Tree-structure based evolutionary algorithms cannot be used for discovering the structure and the parameters of nonlinear system simultaneously, and the evolved solution is usually redundant. This fact also motivates the usage of additive tree approach. Due to the specific computational structure of additive tree, the tree-structure based evolutionary algorithm has been modified so as to make it satisfy few following needs: (1) Different initialization method; (2) Specific instruction sets are used; (3) Different creation and calculation methods of the additive tree; and finally (4) Different evolutionary operators are applied to the additive tree.

5.3.2 Evolving an Optimal or Near-Optimal Structure of Additive Model

Finding an optimal or near-optimal additive tree is an evolutionary process. A number of additive tree variation operators are developed as follows:

Mutation

Four mutation operators were employed to generate offsprings from the parents. These mutation operators are as follows:

(1) Change one terminal node: randomly select one terminal node in the additive tree and replace it with another terminal node;

(2) Change all the terminal nodes: select each and every terminal node in the additive tree and replace it with another terminal node;

(3) Grow: select a random leaf in hidden layer of the additive tree and replace it with a newly generated subtree.

(4) Prone: randomly select a function node in the additive tree and replace it with a terminal node.

Following the work in [259], the additive tree operators were applied to each of the parents to generate an offspring using the following steps: (a) A Poission random number N, with mean λ, was generated. (b) N random mutation operators were uniformly selected with replacement from above four mutation operator set. (c)These N mutation operator were applied in sequence one after the other to the parent to get offspring.

Crossover

Select two additive trees randomly and select one nonterminal node in the hidden layer for each additive tree randomly, and then swap the selected subtree. The crossover operator is implemented with a pre-defined a probability 0.3 in this study.

Selection

EP-style tournament selection [259] with 12 opponents was applied to select the parents for the next generation. This was repeated in each generation for a predefined number of generations or until the best structure is found.

Objective function

Objective function is calculated by the mean square error (MSE) or sum of absolute error(SAE):

$$Fit(i) = \frac{1}{P} \sum_{j=1}^{P} (y_1^j - y_2^j)^2 \tag{5.1}$$

$$Fit(i) = \sum_{j=1}^{P} |y_1^j - y_2^j| \tag{5.2}$$

where P is the total number of training samples, y_1^i and y_2^i are the actual and model outputs of i-th sample. $Fit(i)$ denotes the fitness value of i-th individual;

In order to learn the structure and parameters of the additive tree model simultaneously, a tradeoff between the structure optimization and parameter learning should be taken. In fact, if the structure of the evolved model is not appropriate, it is not useful to pay much attention to the parameter optimization. On the contrary, if the best structure has been found, the further structure optimization may destroy the best structure.

We illustrate a approach for balancing the structure optimization and parameter learning is proposed. If a better structure is found then do random search for a number of steps: maximum allowed steps or stop at case of that no better parameter vector is found for a significantly long time (say 100 to 2000 in our experiments). Where the criterion of better structure is distinguished as follows: if the fitness value of the best program is smaller than the fitness value of the elitist program, or the fitness values of two programs are equal but the nodes of the former is lower than the later, then we say that the better structure is found. Where the best and elitist program is the best program at current generation and the one found so far, respectively.

5.3.3 Parameter Optimization

In the parameter (weights) learning stage, there are a number of learning algorithms (such as GA, EP, gradient based learning) can be used for the tuning of the weights. A random search algorithm [260] is selected only because its simplicity and effectiveness in the aspects of local and global search.

Given a parameter vector $\theta(k) = [\lambda_1(k), \lambda_2(k), \ldots, \lambda_N(k)]$, where k is random search step. Let $x(k) = [x_1(k), x_2(k), \ldots, x_N(k)]$ denotes the small random disturbance vector, which is generated according a probability density function. The random search algorithm can be summarized as follows:

1. Choose an initial value of the parameter vector to be optimized randomly, $\theta(0)$, calculate the objective function, $F(\theta(0))$, and set $k = 0$.
2. Generate random search vector $x(k)$.

   - calculate $F(\theta(k) + x(k))$. If $F(\theta(k) + x(k)) < F(\theta(k))$, the current search is said to be success and then set $y^{(k)} = 1$ and $\theta(k + 1) = \theta(k) + x(k)$. else,
   - calculate $F(\theta(k) - x(k))$. If $F(\theta(k) - x(k)) < F(\theta(k))$, the current search is said to be success too and then set $y^{(k)} = 1$ and $\theta(k + 1) = \theta(k) - x(k)$. otherwise,
   - the search is said to be failure and then set $y^{(k)} = 0$, and

$$\theta(k + 1) = \begin{cases} \theta(k) & If\ K_{er}^+ > K_{er}\ and\ K_{er}^- > K_{er} \\ \theta(k) + x(k) & If\ K_{er}^+ < K_{er}^- \\ \theta(k) - x(k) & If\ K_{er}^+ \geq K_{er}^- \end{cases} \tag{5.3}$$

where $K_{er} \geq 1$ is the maximum error ratio, K_{er}^+ and K_{er}^- are defined by

$$K_{er}^+ = \frac{F(\theta(k) + x(k))}{F(\theta(k))} \tag{5.4}$$

$$K_{er}^- = \frac{F(\theta(k) - x(k))}{F(\theta(k))} \tag{5.5}$$

3. If satisfied solution is found then stop, else set $k = k + 1$ and go to step 2.

It can be seen that the effectiveness of the random search depends largely on the random search vector $x(k)$. Usually for random search the Gaussian probability density functions (PDFs) are used to generate the random search vector [261][262]. In RasID, the used PDF is:

$$f(x_m) = \begin{cases} (1 - q_m)\beta e^{\beta x_m}, if \; x_m \leq 0 \\ q_m\beta e^{-\beta x_m}, if \; x_m > 0 \end{cases} \tag{5.6}$$

where adjustable parameters $q_m \in [0, 1]$ and β are used to control the range and direction of the intensification and the diversification search. Two example graphs of the PDF are shown in Fig. 5.2, from which we can see that the larger the β is, the smaller the local search range is; the larger the q_m is, the higher the search probability in positive direction is. $q_m = 0.5$ means that there is same search probability in positive and in negative direction.

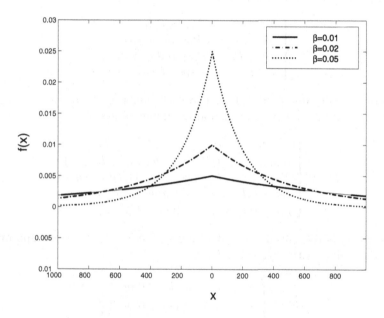

Fig. 5.2 A specific probability distribution function in which the shape of the function depends on the parameters β and q. (a) The larger the β is, the smaller the local search range is. (b) The larger the q_m is, the higher the search probability in positive direction is.

By using the above probability distribution function, the random search vector $x(k)$ can be obtained as follows:

$$x_i(k) = \begin{cases} \frac{1}{\beta}ln\left(\frac{z_i(k)}{1-q_i(k)}\right), & if \; (0 < z_i(k) \leq 1 - q_i(k)) \\ -\frac{1}{\beta}ln\left(\frac{1-z_i(k)}{q_i(k)}\right), & if \; (1 - q_i(k) \leq z_i(k) < 1) \end{cases} \tag{5.7}$$

where $z_i(k)$ is the random real number uniformly distributed at [0,1], $q_i(k) = 0.5$ and $i = 1, 2, \ldots, N$.

The next problem is how to adaptively tune the parameters β and $q_i(k)$ in order to quickly and efficiently find the global minimum in the search space. In the illustrated experiments, the parameter $q_i(k)$ is fixed as $q_i(k) = 0.5$. The parameter β is adaptively changed according to the following equation:

$$\beta = \beta_0 + (\beta_1 - \beta_0) e^{-\phi I_{sf}} \tag{5.8}$$

where ϕ is designed to realize an intensified search and the index I_{sf} for diversified search, β_0 and β_1 are the lower and upper bound of β, respectively.

In addition, the adjustment of the parameters ϕ, I_{sf} and $q_i(k)$ are given in Eq. 5.9, Eq. 5.10 and Eq. 5.11, respectively.

$$\phi = \begin{cases} c_i \phi & P_{sf} > P_{sf0} \\ \phi & P_{sf} = P_{sf0} \ or \ \phi \le \phi_{min} \\ c_d \phi & P_{sf} < P_{sf0} \ and \ \phi > \phi_{min} \\ \phi_0 & k = pre - specified \ integers \end{cases} \tag{5.9}$$

where $c_i \ge 1.0$, $0 < c_d \le 1.0$ are two coefficients assigned with appropriate value, ϕ_0 and ϕ_{min} is the initial and minimum values of ϕ.

$$I_{sf} = \begin{cases} I_{sf0}, & \begin{aligned} & \phi > \phi_{min} \ or \\ & (y^{(k)} = 1 \ and \ I_{sf} > I_{sfmax}) \ or \\ & k = pre - specified \ integers \end{aligned} \\ I_{sf} - \Delta I_{sf1}, & y^{(k)} = 1 \ and \ \phi \le \phi_{min} \\ I_{sf} - \Delta I_{sf2}, & y^{(k)} = 0 \ and \ \phi \le \phi_{min} \end{cases} \tag{5.10}$$

where I_{sf0} is the initial value of I_{sf}, ΔI_{sf1} and ΔI_{sf2} are two appropriate positive values with $\Delta I_{sf1} < \Delta I_{sf2}$.

$$q_i(k) = \begin{cases} \alpha q_i(k), & if \ x_i(k) < 0 \ or \ \frac{\partial^+ F}{\partial \lambda_i} > 0 \\ q_i, & if \ x_i(k) = 0 \ or \ \frac{\partial^+ F}{\partial \lambda_i} = 0 \\ \alpha q_i + (1 + \alpha), & if \ x_i(k) > 0 \ or \ \frac{\partial^+ F}{\partial \lambda_i} < 0 \end{cases} \tag{5.11}$$

where $\alpha \in [0, 1]$ is an appropriate value and $\frac{\partial^+ F}{\partial \lambda_i(k)}$ is the ordered derivative of F for $\lambda_i(k)$.

5.3.4   Summary of General Learning Algorithm

The general learning procedure for the optimal design of additive models can be described as follows:

1) Create an initial population randomly (additive trees and their corresponding parameters);
2) Structure optimization by the additive tree variation operators as described in above

3) If the better structure is found, then go to step 4), otherwise go to step 2);

4) Parameter optimization by random search algorithm as described above. In this stage, the tree structure or architecture of additive model is fixed, and it is the best tree taken from the end of run of the structure search. All the parameters used in the best tree is formulated as a parameter vector to be optimized by local random search algorithm;

5) If the maximum number of local search is reached, or no better parameter vector is found for a significantly long time (100 steps) then go to step 6); otherwise go to step 4);

6) If satisfactory solution is found, then stop; otherwise go to step 2).

5.3.5 Experimental Illustrations

Developed additive models are applied to prediction of chaotic time series, the reconstruction of polynomials and the identification of linear/nonlinear systems.

For each problem, the instruction sets are determined according to the flowing rule: the instruction set I_0 is selected according the complexity estimation of the problem at hard; the instruction set I_1 is selected containing all the terminal instructions and the additional nonterminal instructions $\{R, *_2, \ldots, *_p\}$, here p should not be larger than the input number. In addition, the parameters used for structure evolution are listed in Table 5.1 and the random search algorithm parameters are shown in Table 5.2.

Table 5.1 Parameters used in the flexible neural tree model

Parameter	Initial value
Population Size PS	30
Crossover probability	0.3
Maximum local search steps	2000
Initial connection weights	rand[-1, 1]
Initial parameters a_i and b_i	rand[0,1]

Table 5.2 Parameters used in random search algorithm

β_0	β_1	α	ϕ_0	ϕ_{min}	I_{sf0}	P_{sf0}
0.1	1000	0.995	0.1	0.001	10	0.3
$\triangle I_{sf1}$	$\triangle I_{sf2}$	I_{smax}	K_{er}	c_i	c_d	
0.02	0.1	100	1.001	1.01	0.995	

Example 1: Prediction of Chaotic Time Series

Given the Henon map as follows:

$$x(k+1) = \alpha - x(k)^2 + \beta x(k-1) \qquad (5.12)$$

where $x(k) \in [-2.0, 2.0]$, $\alpha = 1.4$ and $\beta = 0.3$.

100 training data are randomly generated with randomly selecting the initial conditions of $x(0)$ and $x(1)$ by using Eqn. 5.12. The structure and parameters of the system are identified by using the instruction set $I_0 = \{+_2, +_3, +_4, +_5, +_6, +_7, +_8\}$ and $I_1 = \{*, x, R\}$.

The evolved Henon map as the best solution is obtained at generation 127 with fitness 0.007911:

$$x(k+1) = 1.400015 - 1.000050x(k)^2 + 0.300007x(k-1) \qquad (5.13)$$

Figures 5.3 and 5.4 present the outputs of actual system and evolved model and the prediction errors for training data set and validation data set, respectively. It is obvious that the generalization ability of evolved models is very well good, because the evolved model is almost same as the original system model.

In addition, in order to learn about how to select the number of instructions in the instruction set I_0, we varied the instruction set I_0 as follows:

case 1: $I_0 = \{+_2, +_3\}$
case 2: $I_0 = \{+_2, +_3, \cdots, +_5\}$
case 3: $I_0 = \{+_2, +_3, \cdots, +_{10}\}$
case 4: $I_0 = \{+_2, +_3, \cdots, +_{15}\}$

Four independent experiments were done. Simulation results illustrate that a nonlinear system can be identified with a proper selection of instruction set I_0, in which the number of instructions in the instruction set I_0 (the number of nonlinear terms of a nonlinear system to be approximated) will affect the convergence speed of the hybrid method. The smaller the number of instructions is, the faster the convergence speed is. But it is valuable to note that the nonlinear system may not be identified while the number of instructions is too small. The bigger the number of instructions is, the slow the convergence speed is. In our experiments, with the increase of number of instructions, it required 132 generations to get a solution with fitness value 0.028245 for case 3, and 2217 generations with fitness value 0.016810 for case 4.

Example 2: Reconstruction of Polynomials

We try to evolve a more complicated polynomial, a plant to be identified as given by:

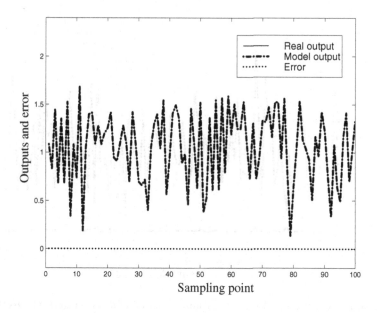

Fig. 5.3 The actual and model outputs for training data set and prediction error

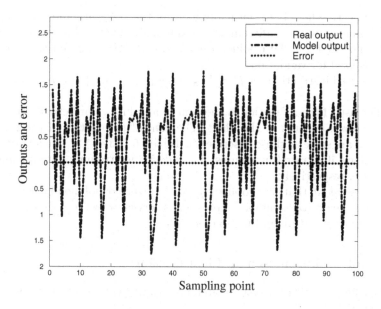

Fig. 5.4 The actual and model outputs for validation data set and prediction error

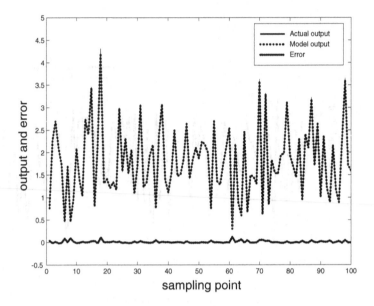

Fig. 5.5 The actual and model outputs for training data set and APPROXIMA-TION error

$$y = 0.2 + 0.3x_1 + 0.4x_2 + 0.5x_3 + 0.6x_1^2 + 0.7x_2^2 + 0.8x_3^2 \qquad (5.14)$$
$$+ 0.9x_1x_2 + 0.1x_1x_3 + 0.2x_2x_3;$$

The objective here is to optimally identify the structure and parameters of the nonlinear system. 400 samples are randomly generated within the interval [0,1]. The first 200 data points are used for the training and the remaining data is used for validation.

The used instruction set $I_0 = \{+_8, +_9, +_{10}, +_{11}, +_{12}\}$ and $I_1 = \{*, x_1, x_2, x_3\}$.

The used fitness function is the sum of absolute error between the actual and evolved outputs of the plant. The control parameters of the proposed method are same as those described in Tables 5.1 and 5.2. The identification results are illustrated in Figures 5.5 and 5.6 for training data set and test data set, respectively.

It is evident that the robustness of the evolved polynomial is very well because the evolved symbolic expression is very close to the actual polynomial in terms of structure and parameter.

Example 3: Linear System Identification

A benchmark ARMAX system often used to test various identification methods is given by [251]:

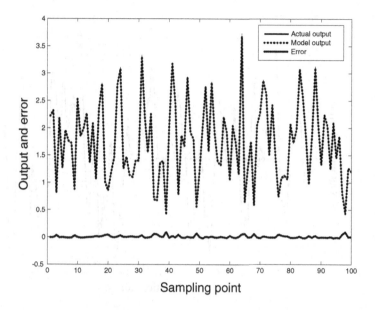

Fig. 5.6 The actual and model outputs for validation data set and approximation error

$$y(k) = b_0u(k-5) + b_1u(k-6) + a_0y(k-1) + a_1y(k-2) \quad (5.15)$$
$$+c_0e(k) + c_1e(k-1) + c_2e(k-2)$$

where $a_0 = 1.5$, $a_1 = -0.7$, $b_0 = 1.0$, $b_1 = 0.5$, $c_0 = 1.0$, $c_1 = -1.0$, $c_2 = 0.2$. The objective here is to optimally identify the structure and parameters of the system in the presence of the noise. The input $u(k)$ is randomly generated at [-5, 5]. The $e(k)$ is Gaussain-distributed random variable with mean 0 and deviation 1. 400 samples are generated by using the above input $u(k)$ and eqn. 5.10, in which 200 data used for training and the other 200 data is used for validation.

The used instruction set is $I_0 = \{+_2, +_3, +_4, +_5, +_6, +_7, +_8, +_9, +_{10}\}$ and $I_1 = \{*, u(k-5), u(k-6), y(k), y(k-1), y(k-2), e(k), e(k-1), e(k-2)\}$.

The used fitness function is the absolute error between the actual and evolved outputs of the plant. The control parameters of the proposed method are shown in Table 5.1. The initial values of parameter in random search are shown in Table 5.2.

The following model is obtained at generation 123 with the fitness 0.669568:

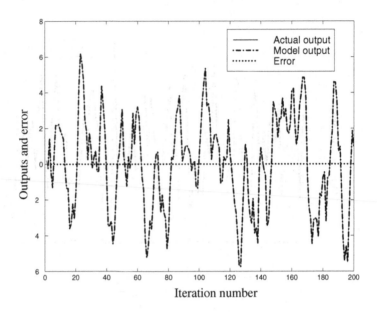

Fig. 5.7 The actual and model outputs for training data set and identification error

$$y(k) = 1.000307u(k-5) + 0.505093u(k-6) \qquad (5.16)$$
$$+1.499481y(k-1) - 0.699698y(k-2)$$
$$+0.998582e(k) - 1.001312e(k-1)$$
$$+0.198232e(k-2)$$

Figure 5.7 presents the outputs of actual system and evolved model and identification error training data set. The generalization ability of the model is illustrated in Figure 5.8.

Example 4: Nonlinear System Identification

A second-order non-minimum phase system with gain 1, time constants $4s$ and $10s$, a zero at $1/4s$, and output feedback with a parabolic nonlinearity is chosen to be identified. With sampling time $T_0 = 1s$, this system follows the nonlinear difference equation:

$$y(k) = -0.07289[u(k-1) - 0.2y^2(k-1)] + 0.09394[u(k-2) \qquad (5.17)$$
$$-0.2y^2(k-2)] + 1.68364y(k-1) - 0.70469y(k-2).$$

where the input lie in the interval [-1,1].

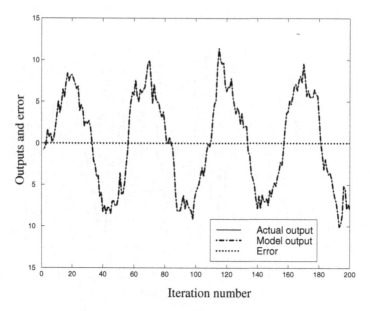

Fig. 5.8 The actual and model outputs for validation data set and identification error

The training data and test data are generated using the input signals shown in Figures 5.9 and 5.10. The used instruction sets are $I_0 = \{+_3, \ldots, +_8\}$, $I_1 = \{*, u(k-1), u(k-2), y(k-1), y(k-2)\}$. The evolved model is almost same as the original system. The used cost function is the sum of the absolute error (SAE). The SAE for training data and test data are 1.873652 and 2.349801, respectively. A comparison between process and simulated model outputs on training data set is depicted in Figure 5.11. Figure 5.12 presents the model and system outputs and the identification error for validation data set.

From above simulation results, it can be seen that the proposed method works very well for generating system models.

5.3.6 Discussions

We illustrated the design of an additive tree model and an optimization algorithm. In the viewpoint of the calculation for structure, the additive tree model can be seen as a natural computational structure for modeling a class of nonlinear systems, in which the characteristics of nonlinear systems can be reconstructed and captured by using automatically evolved additive tree model.

Simulation results for the prediction of chaotic time series, the reconstruction of polynomials and the identification of linear/nonlinear systems show the feasibility and effectiveness of the proposed method. The evolved additive tree models have following two advantages: (1) it is robust, and (2) it is

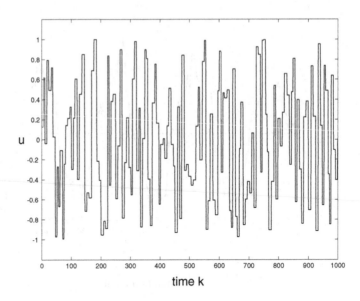

Fig. 5.9 Excitation input signals for generating the training data set

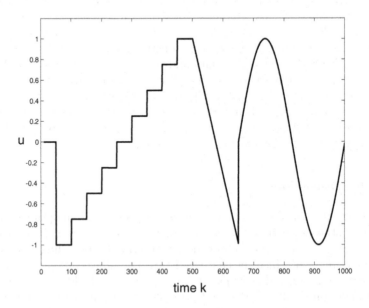

Fig. 5.10 Excitation input signals for generating the validation data set

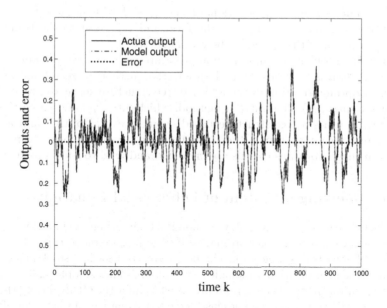

Fig. 5.11 Comparison between process and simulated model output for the training data set

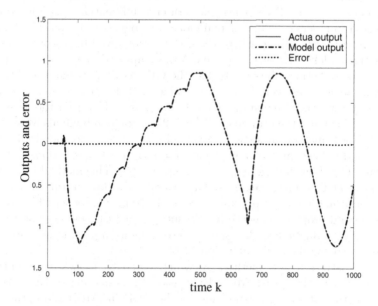

Fig. 5.12 Comparison between process and simulated model output for the validation data set

easy to analysis by using traditional techniques. This is because the evolved additive tree model is simple in the form and very near to the traditional representation of the system to be reconstructed.

The key problem for finding an appropriate additive tree to model a non-linear system is that how to find an effective method to search an optimal or near-optimal solution in the additive tree structure space and related parameter space. We illustrated a method, which alternatively search the two spaces by using a tree-structure based evolutionary algorithm and a local random search algorithm. Other tree-structure based methods and parameter learning algorithms can also be modified and used to solve problems.

5.4   Inferring a System of Differential Equations

In the fields of physics, chemistry, economics, bioinformatics etc., a lot of problems can be expressed by the ordinary differential equations(ODEs). The problems of forecasting, quantum mechanics, wave propagation, stock market dynamics and identification of biological systems are some examples [263]. For this reason various methods have been proposed to infer the ODEs during the last few years. The researches can be classified into two classes: one is to identify the ordinary differential equation's parameters and the other is to identify the ordinary differential equation's structure. The former is exemplified by the Genetic Algorithms (GA), and the latter by the Genetic Programming (GP).

Cao et al. used GP to evolve the ordinary differential equations from the observed time series [264]. The main idea was to embed the genetic algorithm in genetic programming, where the GP was employed to discover and optimize the model's structure, and the GA was employed to optimize the model's parameters. Authors illustrated that the GP-based approach introduced numerous advantages over the most available modeling methods. Iba proposed ordinary differential equations identifying method based on the least mean square(LMS) and the ordinary GP [265][273]. Some individuals were created by the LMS method at some intervals of generations and they replaced the worst individuals in the population. Tsoulos and Lagar proposed a novel method based on the grammatical evolution [263]. This method formed generations of trial solutions expressed in an analytical closed form. The Bayesian inferential methodology provides a coherent framework with which to characterize and propagate uncertainty in such mechanistic models and this provides an introduction to Bayesian methodology as applied to system models represented as differential equations [274].

We illustrate a new method, in which the particle swarm optimization (PSO) is used along with Multi Expression Programming (MEP). We infer the structure of the right-hand sides of the ODEs by MEP and optimize the parameters of the ordinary differential equations by the PSO algorithm. The partitioning [284] is used in the process of identifying the system's structure. Each ordinary differential equation of the ordinary differential equations can be inferred separately and the search space rapidly reduces.

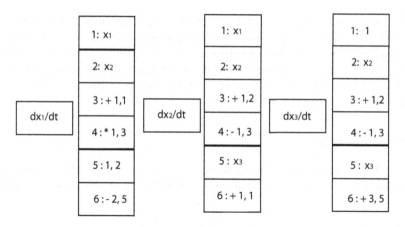

Fig. 5.13 Example of a ordinary differential equations

5.5 Inference of Differential Equation Models by Multi Expression Programming

5.5.1 Structure Optimization by the MEP

Encoding

Multi Expression Programming (MEP) is a variant of the genetic programming, which is proposed by Oltean in 2002 [266] [267]. The traditional GP [268] encodes a single expression (computer program). However, the MEP chromosome encodes several genes. Each gene encodes a terminal or a function symbol, which is selected from a terminal set T or a function set F. The two sets for a given problem are pre-defined. A gene that encodes a function includes some pointers towards the function arguments. The number of the pointers depends on how many arguments the function has. The best encoded solution is chosen to represent the chromosome [269].

MEP is used to identify the form of the system's differential equations. For this purpose, we encode the right-hand side of each ordinary differential equation into a MEP chromosome. For example aN ordinary differential equations model with the form of:

$$\begin{cases} \dot{X}_1 = aX_1 + bX_2 \\ \dot{X}_2 = cX_1 \\ \dot{X}_3 = dX_2 + e \end{cases} \tag{5.18}$$

can be represented as three MEP chromosomes $\{E_3, E_6, E_3\}$ illustrated in Figure 5.13, where the coefficients a, b, c, d, e are derived by the PSO (described later in this Section).

We infer the system of ordinary differential equations using partitioning. A Partition in which the equations describing each variable of the system can be inferred separately, and thus significantly reducing the research space. When using partitioning, a candidate equation for a sigle variable is integrated by substituting references to other variables with the data from the observed time series [284].

Fitness Function

The MEP chromosome contains some expressions, so each expression E_i is calculated by the root mean squared error(RMSE) or the sum of absolute error(SAM):

$$f(E_i) = \sqrt{\frac{1}{n} \sum_{j=1}^{n} (x_{ji} - x'_j)} \tag{5.19}$$

$$f(E_i) = \sum_{j=1}^{n} \left| x_{ji} - x'_j \right| \tag{5.20}$$

where x_{ji} is the time series by expression E_i and x'_j is the targeted time series. The fitness of a chromosome is equal to the best fitness of the expressions encoded.

Genetic Operators

The genetic operators used in the MEP algorithm are crossover and mutation [266].

(1) Crossover. We choose the one-point crossover. Two parents are selected according to the predefined crossover probability P_c. One crossover point is randomly chosen and the parents exchange the sequences at this point.
(2) Mutation. One parent is selected according to the predefined mutation probability P_m. One mutation point is randomly chosen. If the mutation position encodes a function symbol, it may be mutated into a terminal symbol or another function with arguments and parameters. And we can mutate the function arguments and parameters into random arguments and parameters.

5.5.2   Parameter Optimization by Particle Swarm Optimization Algorithm

Encoding

At the beginning of this process, we check all the constants contained in each equation , namely we count their number n_i and report their places. The distribution of parameters in each chromosome is illustrated in Fig. 5.14.

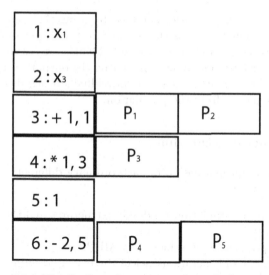

Fig. 5.14 Distribution of parameters in each chromosome.

According to n_i, the particles are randomly generated initially. Each particle x_i represents a potential solution. A swarm of particles moves through space, with the moving velocity of each particle represented by a velocity vector v_i. At each step, each particle is evaluated and keeps track of its own best position, which is associated with the best fitness that has achieved so far in a vector $Pbest_i$. And the best position among all the particles is kept as Gbest [270]. A new velocity for particle i is updated by

$$v_i(t+1) = v_i(t) + c_1 r_1 (Pbest_i - x_i(t)) + c_2 r_2 (Gbest(t) - x_i(t)) \qquad (5.21)$$

where c_1 and c_2 are positive constant and r_1 and r_2 are uniformly distributed random number in $[0,1]$. Based on the updated velocities, each particle changes its position according to the following equation:

$$x_i(t+1) = x_i(t) + v_i(t+1) \qquad (5.22)$$

5.5.3   Fitness Definition

The fitness of each variable is defined as the sum of squared error and the penalty for the degree of the equations:

$$fitness(i) = \sum_{k=0}^{T-1} (x_i'(t_0 + k\Delta t) - x_i(t_0 + k\Delta t))^2 + a \qquad (5.23)$$

where t_0 is the starting time, Δt is the step size, T is the number of the data point, $x_i(t_0 + k\Delta t)$ is the actual outputs of i-th sample, and $x_i'(t_0 + k\Delta t)$ is the

ODEs' outputs. All outputs are calculated by using the approximate forth-order Runge-Kutta method. a is the penalty for the degree of the equations. To reduce the problem search space, the individuals' degrees are limited to the stated range according to a. When calculating the outputs, some individuals may cause overflow. In such cases, the individual's fitness becomes so large that it will be removed from the population.

5.5.4   Summary of Algorithm

The MEP for the optimal design of each ordinary differential equation can be described as follows:

(1) Create an initial population randomly (structures and their correspond-
    ing parameters);
(2) Structure optimization is achieved by MEP;
(3) At some interval of generations, select the better structures and optimize
    its parameters. Parameter optimization is achieved by the PSO. In this
    process, the structure is fixed.
(4) If satisfactory solution is found, then stop; otherwise go to step (2).

If the ordinary differential equations' parameters have some error, we can use the standard fourth-order Runge-Kutta method to integrate the ordinary differential equations to optimize parameters.

5.6   Modeling Chemical Reactions

We have prepared four tasks to test the effectiveness of the method. Experimental parameters are summarized in Table 5.3. Function and terminal sets F and T are follows:

$$F = \{+, -, *\}$$
$$T = \{X_1, ..., X_n, 1\}. \tag{5.24}$$

Table 5.3 Parameters for experiments

	Exp1	Exp2	Exp3
Population size	20	50	50
Generation	50	100	100
Crossover rate	0.7	0.7	0.7
Mutation rate	0.3	0.3	0.3
Time series	1	1	1
Stepsize	0.01	0.05	0.05
Data point	30	30	48

5.6.1 Simple Chemical Reaction Model

The reaction equations [264] are described below:

$$HCHO + (NH_2)_2CO \xrightarrow{k_1} H_2N \cdot CO \cdot NH \cdot CH_2OH \qquad (5.25)$$

$$H_2N \cdot CO \cdot NH \cdot CH_2OH + (NH_2)_2CO \xrightarrow{k_2} (NH_2CONH)_2CH_2 \qquad (5.26)$$

As a kind of typical consecutive reactions, the concentrations of the three components in the system satisfy the following system:

$$\begin{cases} \dot{X}_1 = -1.4000X_1 \\ \dot{X}_2 = 1.4000X_1 - 4.2X_2 \\ \dot{X}_3 = 4.2000X_2 \end{cases} \qquad (5.27)$$

The time series were generated for the above set of reactions with initial

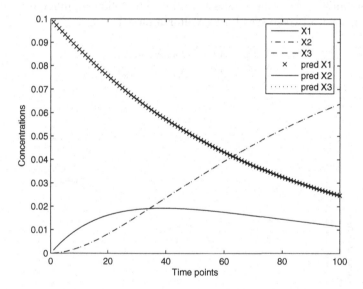

Fig. 5.15 Time series of the acquired model for chemical reaction

conditions $\{0.1, 0, 0\}$ for $\{X_1, X_2, X_3\}$. Experimental parameter for this task are depicted in Table 1. We have acquired the system of eq. (5.28), which gave the sums of sums of absolute errors as $(X_1, X_2, X_3) = (3.6 \times 10^{-12}, 4.01 \times 10^{-12}, 8.79 \times 10^{-12})$. The time series generated is depicted in Figure 5.15 along with that of the target.

$$\begin{cases} \dot{X_1} = -1.400017X_1 \\ \dot{X_2} = 1.400044X_1 - 4.199987X_2 \\ \dot{X_3} = 4.199939X_2 \end{cases} \qquad (5.28)$$

The best kinetic model acquired in [264] was as follows:

$$\begin{cases} \dot{X_1} = -1.400035X_1 \\ \dot{X_2} = 1.355543(X_1 + t) - 4.482911X_2 \\ \dot{X_3} = 4.069420X_2 + t - 0.002812 \end{cases} \qquad (5.29)$$

where the sums of squared errors were $(X_1, X_2, X_3) = (1.6 \times 10^{-11}, 3.24 \times 10^{-8}, 3.025 \times 10^{-9})$. Note that the terminal set in [264] included the time variable t.

5.6.2   Two-Species Lotka-Volterra Model

The Lotka-Volterra model describes interactions between two species, i.e., predators and preys, in an ecosystem [271]. The following differential equations represent a two-species Lotka-Volterra model:

$$\begin{cases} \dot{X_1} = 3X_1 - 2X_1X_2 - X_1{}^2 \\ \dot{X_2} = 2X_2 - X_1X_2 - X_2{}^2 \end{cases} \qquad (5.30)$$

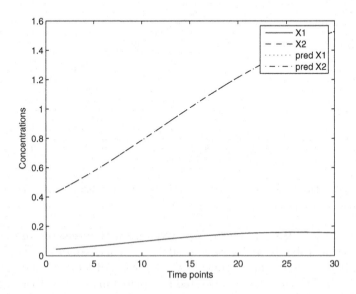

Fig. 5.16 Time series of the acquired model for Lotka-Volterra model

A time series was generated for the above set of reactions with initial conditions $\{0.04, 0.4\}$ for $\{X_1, X_2\}$. The time series generated is shown in Figure 5.16. Experimental parameter settings for this task are shown in Table 5.3. We have acquired the system of eq. (5.31), which gave the sums of sums of absolute errors as: $(X_1, X_2)=(2.5 \times 10^{-11}, 4.45 \times 10^{-10})$. In all runs,we have succeeded in getting almost the same differential equations.

$$\begin{cases} \dot{X}_1 = 2.999998X_1 - 2.000081X_1X_2 - 0.9993\,X_1{}^2 \\ \dot{X}_2 = 2.000005X_2 - 1.000064X_1X_2 - 0.999997X_2{}^2 \end{cases} \tag{5.31}$$

The best model acquired in [284] was eq.(5.32). When compared to that, structure is the same and the parameters of obtained using the proposed model are closer to the target model.

$$\begin{cases} \dot{X}_1 = 3.0014X_1 - 2X_1X_2 - X_1{}^2 \\ \dot{X}_2 = 2.0001X_2 - X_1X_2 - X_2{}^2 \end{cases} \tag{5.32}$$

5.6.3   Bimolecular Reaction

The bimolecular reaction equations [272] are described below:

$$X_2 + X_1 \xrightarrow{k_1} X_3 \tag{5.33}$$

$$X_3 \xrightarrow{k_2} X_4 + X_2 \tag{5.34}$$

The corresponding rate equations for all the four species are as follows:

$$\begin{cases} \dot{X}_1 = -2X_1X_2 \\ \dot{X}_2 = -2X_1X_2 + 1.2X_3 \\ \dot{X}_3 = 2X_1X_2 - 1.2X_3 \\ \dot{X}_4 = 1.2X_3 \end{cases} \tag{5.35}$$

The time series were generated for the above set of reactions with initial conditions$\{1, 0.1, 0, 0\}$for $\{X_1, X_2, X_3, X_4\}$, which is shown in Figure 5.17 along with the target time series. Experimental parameter for this task are illustrated in Table 5.3.

We have acquired the system of eq. (5.37), which gave the sums of absolute errors as $(X_1, X_2, X_3, X_4)=(\,1.6\times10^{-11}, 9.0\times10^{-12}, 8.8\times10^{-12}, 2.5\times10^{-11})$.

$$\begin{cases} \dot{X}_1 = -1.9920X_1X_2 \\ \dot{X}_2 = -1.1983X_1X_2 + 1.9920X_3 \\ \dot{X}_3 = 1.9920X_1X_2 - 1.1983X_3 \\ \dot{X}_4 = 1.1983X_3 \end{cases} \tag{5.36}$$

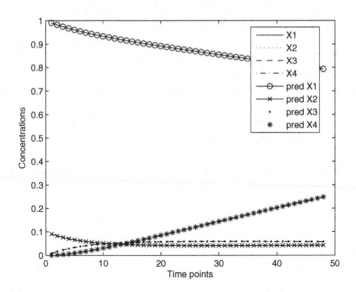

Fig. 5.17 Time series of the acquired model for bimolecular reaction

Compared with eq. (5.36) [272], the developed model and parameters are closer to the target system and the model is able to predict the standard enzyme kinetics scheme with the rate parameters closer to the generative values.

$$
\begin{cases}
\dot{X}_1 = -1.99999 X_1 X_2 \\
\dot{X}_2 = 1.20000 X_3 - 1.99999 X_1 X_2 \\
\dot{X}_3 = -1.20000 X_3 - 2.00000 X_1 X_2 \\
\dot{X}_4 t = 1.99999 X_3
\end{cases}
\tag{5.37}
$$

5.7   Inferring Gene Regulatory Networks

Gene expression programs, which produce the living cells involving regulated transcription of thousands of genes depend on recognition of specific promoter sequences by transcriptional regulatory proteins. The problem is how a collection of regulatory proteins associates with genes can be described as a transcriptional regulatory network. The most important step is to identify the interactions among genes by the modeling of gene regulatory networks.

Many models have been proposed to describe the network including the Boolean network [276][282], Dynamic Bayesian network [277], the system of differential equations [278] and so on. A recent review for inferring genetic regulatory networks based on data integration and dynamical models is available in [283]. The system of differential equations is powerful and flexible model to describe complex relations among components [280], so many methods are proposed for inferring a system of differential equations for the gene regulatory network during the last few years. But it is hard to determine the suitable form of equations, which describe the network. In the previous studies, the form of the differential equation was being fixed. The only one goal was to optimize parameters and coefficients. For example, Tominaga used Genetic Algorithms (GA) to optimize the parameters of the fixed form of system of differential equations [279]. In recent years some researchers studied the learning of gene regulatory network by inferring the structures and parameters of a system of ODEs. Sakamoto proposed an ODEs identification method by using the least mean square (LMS) along with the ordinary genetic programming (GP) to identifying the gene regulatory network [280]. Cho proposed a new representation named S-tree based GP to identify the structure of a gene regulatory network and to estimate the corresponding parameter values at the same time [281]. Li proposed GP was applied to identify the structure of model and Kalman filtering was used to estimate the parameters in each iteration. Both standard and robust Kalman filtering were considered [275]. But their inference algorithms can only be applied to the small-scale networks.

The form of the ODE is usually represented by:

$$\dot{X}_i = f_i(X_1, X_2, ..., X_n)(i = 1, 2, ..., n) \tag{5.38}$$

where X_i is the state variable and n is the number of components in the gene regulatory network. In a network, X_i is the expression level of the ith gene and n is the number of genes in the network.

MEP is used to identify the form of the system of differential equations. For this purpose, we encode right-hand side of each ODE into a MEP chromosome. We infer the system of ODEs with partitioning. Partition, in which equations describing each variable of the system can be inferred separately, significantly reducing the search space. When using partitioning, a candidate equation for a signal variable is integrated by substituting references to other variables with data from the observed time series. This allows us to infer the structure of systems comprising more variables and higher degree of coupling than were inferred by other methods [284].

The best ODEs is comprised of the equation obtained in each process. If the parameters of ODEs have some error, we can use the standard fourth-order Runge-Kutta method to integrate the ODE to optimize parameters.

We have prepared two tasks to test the effectiveness of the method. Experimental parameters are summarized in Table 5.4. Function and terminal sets F and T are described as follows:

$$F = \{+, -, *, x^a\}$$
$$T = \{X_1, ..., X_n, 1\}. \tag{5.39}$$

Table 5.4 Parameters for experiments

	Exp1	Exp2
Population size	5000	1000
Generation	2000	2000
Crossover rate	0.7	0.7
Mutation rate	0.3	0.3
Time series	1	1
Stepsize	0.01	0.05
Data point	15	20
gene size	5	15
PSO Population size	100	100
PSO Generation	100	100

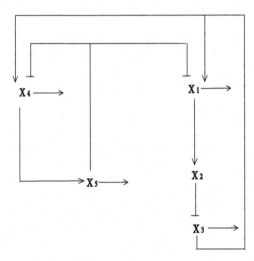

Fig. 5.18 The targeted gene regulator network

5.7.1   The Small Artificial Gene Regulatory Network

Figure 5.18 illustrates an example of gene regulatory network. This type of network can be modeled as a so-called S-system model [286]. This model is based on approximating kinetic laws with multivariate power-law functions. A model consists of n non-linear ODEs and the generic form of equation i is given as follows:

$$X_i'(t) = \alpha_i \prod_{j=1}^{n} X_j^{g_{ij}}(t) - \beta_i \prod_{j=1}^{n} X_j^{h_{ij}}(t) \tag{5.40}$$

Table 5.5 Parameters of the genetic network system

i	α_i	g_{i1}	g_{i2}	g_{i3}	g_{i4}	g_{i5}	β_i	h_{i1}	h_{i2}	h_{i3}	h_{i4}	h_{i5}
1	5.0		1.0		-1.0		10.0	2.0				
2	10.0	2.0					10.0		2.0			
3	10.0		-1.0				10.0		-1.0	2.0		
4	8.0			2.0		-1.0	10.0				2.0	
5	10.0				2.0		10.0					2.0

where X is a vector of dependent variable, α and β are vectors of non-negative rate constants and g and h are matric of kinetic orders.

The parameter of the genetic network are given in Table 5.5 and the initial conditions are $\{0.7, 0.12, 0.14, 0.16, 0.18\}$ for X_1, X_2, X_3, X_4, X_5 [285]. Experimental parameter settings for this task are shown in Table 5.4. The search region of the parameters was $[0.0, 15.0]$. Five runs are carried out. In each run, the proposed method produces one candidate solution. Selecting 30 better structures to optimize parameters by PSO at every 10 generations. To handle the powers of the component variable, we used the following terminal set:

$$T = \{X_1, X_1^{-1}, X_2, X_2^{-1}, X_3, X_3^{-1}, X_4, X_4^{-1}, X_5, X_5^{-1}\} \tag{5.41}$$

We created the following ODEs by the method and throughout the simulations, we further confirm that the identified system is quite close to the original system (Fig. 5.19).

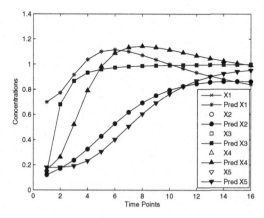

Fig. 5.19 Time series of the acquired model

Table 5.6 Obtained Parameters of the ODEs by the proposed method and S-tree(GP), α_i', β_i':parameters by the proposed method, α_i'', β_i'': parameters by S-tree(GP), α_i, β_i:true parameters

i	$\alpha_i'/\alpha_i''/\alpha_i$	$\beta_i'/\beta_i''/\beta_i$
1	8.5854/4.9999/5.0	13.7959/9.9999/10.0
2	9.7709/10.0000/10.0	10.0117/10.0000/10.0
3	13.7629/10.0000/10.0	13.9742/10.0000/10.0
4	8.3954/8.0000/5.0	13.7959/10.0000/10.0
5	9.4643/9.9999/5.0	13.7959/10.0000/10.0

$$
\begin{cases}
\dot{X}_1 = 4.999994 X_3 X_5^{-1} - 9.999994 X_1^2 \\
\dot{X}_2 = 10.000023 X_1^2 - 10.000014 X_2^2 \\
\dot{X}_3 = 10.000016 X_2^{-1} - 10.000015 X_3^2 X_2^{-1} \\
\dot{X}_4 = 8.000003 X_3^2 X_4 - 10.000001 X_3^2 \\
\dot{X}_5 = 9.999994 X_4^2 - 10.000019 X_5^2
\end{cases}
\qquad (5.42)
$$

Cho [281] proposed a new representation named S-tree based genetic programming(GP) to identify the structure of a gene regulatory network and the size of population was assumed as 10,000 and the proposed scheme was terminated after 5×10^5 iterations. Compared to that, for the proposed method, size of population and number of iterations are far smaller (Table 5.5). We also obtained the true structure during every experiment. Table 5.6 depicts the best parameters obtained among all the experiments. Obviously the parameters are very closer to the target model.

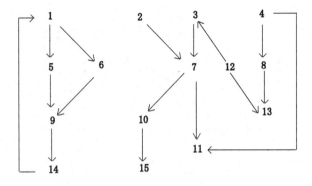

Fig. 5.20 The large-scale Artificial Gene Regulatory Network

Table 5.7 S-system parameters of the large-scale target model of the large artificial gene regulatory network

α_i	1.0
β_i	1.0
$g_{i,j}$	$g_{1,14} = -0.1, g_{3,12} = -0.2, g_{5,1} = 1.0,$
	$g_{6,1} = 1.0, g_{7,2} = 0.5, g_{7,3} = 0.4, g_{8,4} = 0.2,$
	$g_{9,5} = 1.0, g_{9,6} = -0.1, g_{10,7} = 0.3, g_{11,4} = 0.4,$
	$g_{11,7} = -0.2, g_{12,13} = 0.5, g_{13,8} = 0.6, g_{14,9} = 1.0,$
	$g_{14,15} = -0.2, g_{15,10} = 0.2, otherg(i,j) = 0.0$
$h_{i,j}$	1.0 if $i = j$, 0.0 otherwise

5.7.2 The Large-Scale Artificial Gene Regulatory Network with Noisy Environment

This test system, which is the same with Experiment 1 is a reduced version of test system ss30 genes that was introduced by [287]. Figure 5.20 illustrates the example of gene regulatory network. The system represents a genetic network with 15 variables. Table 5.7 shows the parameters of S-system formalism.

As we have to estimate a relatively large number of parameters and structure of the system of differential equations with a small data set, there can be a lot of different possible network structures all of which bring about only small differences in estimating the given data set. These false candidates can be decreased by reducing the structural search space based on the available constraint. The constant is that all the diagonal elements in the matrix h are not zero (h_{ii} for $i = 1, ..., n$) [281]. Namely the i-th equation must contain X_i. This is because as the concentration X_i is higher, X_i can participate in the reaction more actively (i.e. it disappears fast) [281].

The set of time-series data began from randomly generated initial values and was obtained by solving the set of differential equations of the targeted model. In the past, for large-scale artificial gene regulatory network, the form of the differential equation was fixed, and the only one goal was to optimize the parameters and coefficients [288]. We illustrate the Multi Expression Programming approach to evolve the right hand side of the equation. Experimental parameters for this task are shown in Table 5.4. Ten runs were carried out and the search region of the parameters was [-1.0, 1.0]. During the experiments, we could obtain the best structure and parameters which were the same with the target model (Figure 5.20 and Table 5.7) except the 11-th gene. We only obtained the 11-th differential equation: $\dot{X}_{11} = X_7^{-0.199999} X_{11}^{0.018851} - X_{11}$. So the only interaction $X_4 \rightarrow X_{11}$ could not be identified.

To test the performance of the proposed method in a real-world setting, we added 1, 2, 5, 10 and 15% Gaussian noise to the time-series data in order to simulate the measurement noise that often corrupts the observed data obtained from actual measurements of gene expression patterns. Except that the size of population is fixed at 10000, other settings are same as in the previous experiment. In the same execution time for each run, we obtained the same structure by the time-series data, which has added 1, 2, 5 and 10% Gaussian noise as the data without noise. When the noise ratio is similar up to the 15%, the $X_3 \to X_7$ and $X_4 \to X_{11}$ are not identified. Hence, we can conclude that the proposed algorithm is robust within 10% random noise.

Not having the fixed form, the proposed method can automatically identify the structure and parameters of a network. In general the networks have sparse structures. Without fixed form many irrelevant parameters need not be optimized. Thus the computational complexity reduces largely. Time-series data should be chosen carefully since an improper value can impair the success rate of attaining good candidates for a true network structure [289].

By several experiments, we succeeded in creating the systems of differential equations, which are close to the target systems. The results show the effectiveness and veracity of the proposed method. The method has following two advantages: (1) a MEP chromosome encodes several expressions, so we can acquire the best structure of the ordinary differential equations only by a small population; (2) by partitioning, we can acquire the best system very fast.

Part VI

Conclusions and Future Research

6

Concluding Remarks and Further Research

6.1 Limitations of Conventional Computational Intelligence

Real-world problems are typically ill-defined systems, difficult to model and with large-scale solution spaces. In these cases, precise models are impractical, too expensive, or non-existent. The relevant available information is usually in the form of empirical prior knowledge and input-output data representing instances of the system's behavior [290]. Soft Computing (SC), including Neural Computing (NC), Fuzzy Computing (FC), Evolutionary Computing (EC) etc., provides us with a set of flexible intelligent computational tools to perform approximate reasoning, learning from data, search tasks etc.

NC, FC, EC, among others, have been established and illustrated their strength and drawbacks. NC can perform ideally in domains of purely numerical nature, as well as in making effective predictions in time series data and nonlinear function approximations. EC could competitively perform optimization tasks in a very large search space, identifying sub-optimal solutions of high quality, becoming thus the methods of choice for domains suffering from combinatorial explosion phenomena such as operations research, manufacturing etc. FC has been provide ideal for handling approximate concepts, human characterizations and domains having unclear boundaries. Moreover, it has been observed that the highly increasing computing power and technology, could make possible the use of more complex intelligent architectures, taking advantage of more than one intelligent techniques, not in a competitive, but rather in a collaborative sense. Therefore, discovering of more sophisticated and new evolutionary learning models and its application to new areas and problems still remain as key questions for the next 10 years.

Y. Chen, A. Abraham.: Tree-Struc. Based Hybrid Com. Intelligence, ISRL 2, pp. 185–190.
springerlink.com   © Springer-Verlag Berlin Heidelberg 2010

6.2   Towards Tree-Structure Based Hierarchical Hybrid Computational Intelligence

To investigate the hybrid technique further, an Hierarchical Hybrid Computational Intelligence (HHCI) framework is illustrated in this book. Based on tree-structure based encoding and the specific function operators, the new HHCI models can be flexibly constructed and evolved by using simple computational intelligence techniques.

6.2.1   Tree Structure Based Evolutionary Computation Models

Tree-structure based evolutionary algorithms including Genetic Programming (GP), Probabilistic Incremental Program Evolution (PIPE) and the recent variants of GP, i.e., Gene Expression Evolution (GEP), Estimation of Distribution Programming (EDP) and Multi Expression Evolution (MEP), have been an active are of research in recent years.

Motivated by hierarchical fuzzy systems, a natural extension of traditional computational intelligence (CI) models is to introduce some intermediate levels of processing so that the HHCI models can be constructed. In this perspective, existing CI components should be employed to formulate different HHCI models and the architecture of HHCI model could be determined by the hierarchical nature and its learning ability of tree-structure based evolutionary algorithms.

6.2.2   Hierarchical Hybrid Computational Intelligence Framework

A tree-structure based HHCI framework is illustrated in Figure 6.1(e). A function operator set $\{T_2, T_3, \ldots, T_n\}$ and a terminal set $\{x_1, x_2, \ldots, x_n\}$ are employed in the construction of the tree. The function operator $T_i(i = 2, 3, \ldots, n)$ denotes that the node has i arguments with the operator type of T. Some possible types of the function operators are described as follows.

Types of Function Operators

Flexible Neuron Operator. The flexible neuron operator is illustrated in Figure 6.1(a). Assume that the used flexible activation function is:

$$f(a_i, b_i, x) = exp(-(\frac{x - a_i}{b_i})^2). \tag{6.1}$$

where a_i, b_i are free parameters. The output the operator can be calculated as:

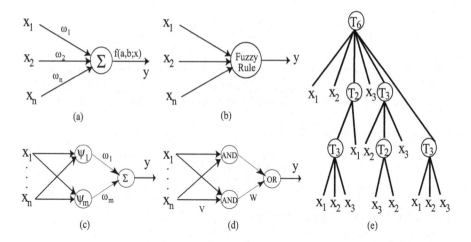

Fig. 6.1 Some of the function operators and an HHCI tree. (a)a flexible neuron operator, (b)Takagi Sugeno fuzzy operator, (c)basis function operator, (d)a fuzzy neural operator, and (e)a general representation (encoding) of HHCI model, where the used function set is $F = \{T_2, T_3, \ldots, T_6\}$ and the terminal set is $T = \{x_1, x_2, x_3\}$.

$$y = exp(-(\frac{\sum_{j=1}^{n} \omega_j x_j - a_i}{b_i})^2). \tag{6.2}$$

where ω_j denotes adjustable connection strength of node and sub-node.

TS Fuzzy Operator. The fuzzy operator is shown in Figure 6.1(b). Takagi and Sugeno developed a hybrid modelling technique designed to combine conventional and fuzzy modelling. The Takagi Sugeno models are represented by a series of fuzzy rules of the form:

$$R_i: \text{IF}(x \text{ is } A_i) \text{ THEN } (y = f_i(x_i))$$

where $f_i(x)$ is a local model used to approximate the response of the system in the region of the input space represented by the antecedent. The overall model output is calculated as the normalized sum:

$$y = \frac{\sum_{i=1}^{p} \mu_{A_i}(x) f_i(x_i)}{\sum_{k=1}^{p} \mu_{A_k}(x)} \tag{6.3}$$

where normalized fuzzy membership functions determine which local models are valid given a particular input.

Basis Function Operator. The basis function operator is shown in Figure 6.1(c). In general, the basis function networks can be represented as:

$$y = \sum_{i=1}^{m} \omega_i \psi_i(x; \theta) \tag{6.4}$$

where $x \in R^n$ is input vector, $\psi_i(x; \theta)$ is ith basis function, and ω_i is the corresponding weights of ith basis function and θ is the parameter vector used in the basis functions. Examples of typical basis functions include:

(a) Gaussian radial basis function

$$\psi_i(x; \theta) = \prod_{j=1}^{n} exp(-\frac{\| x_j - b_j \|^2}{a_j{}^2}) \tag{6.5}$$

(b) Order 2 B-Spline basis function

$$\psi_i(x; \theta) = \prod_{j=1}^{n} B_2(\frac{x_j - b_j}{a_j}) \tag{6.6}$$

where the translation and dilation of the order 2 B-spline function is given by:

$$B_2(\frac{t-b}{a}) = \begin{cases} \frac{9}{8} + \frac{3}{2}(\frac{t-b}{a}) + \frac{1}{2}(\frac{t-b}{a})^2, & t \in [-\frac{3}{2}a + b, -\frac{1}{2}a + b) \\ \frac{3}{4} - (\frac{t-b}{a})^2, & t \in [-\frac{1}{2}a + b, \frac{1}{2}a + b) \\ \frac{9}{8} - \frac{3}{2}(\frac{t-b}{a}) + \frac{1}{2}(\frac{t-b}{a})^2, & t \in [\frac{1}{2}a + b, \frac{3}{2}a + b] \\ 0, & otherwise \end{cases} \tag{6.7}$$

(c) Wavelet basis function

$$\psi_i(x; \theta) = \prod_{j=1}^{n} \frac{1}{\sqrt{|a_j|}} \phi(\frac{x_j - b_j}{a_j}) \tag{6.8}$$

where ϕ is a mother wavelet, i.e., a Mexican Hat: $\phi(t) = \frac{2}{\sqrt{3}}\pi^{-1/4}(1 - t^2)e^{-t^2/2}$.

Fuzzy Neural Operator. The fuzzy neural operator is illustrated in Figure 6.1(d). AND neuron is a nonlinear logic processing element with n-inputs $x \in [0, 1]^n$ producing an output y_1 governed by the expression:

$$y_1 = AND(x; w) \tag{6.9}$$

where w denotes an n-dimensional vector of adjustable connections (weights). The composition of x and w is realized by an t-s composition operator based on $t-$ and $s-$norms, that is:

$$y_1 = T_{i=1}^{n}(w_i S x_i) \tag{6.10}$$

with s denoting some s-norm and t standing for a t-norm. By reverting the order of the t- and s-norms in the aggregation of the inputs, we end up with a category of OR neurons:

$$y_2 = OR(x; w) \tag{6.11}$$

$$y_2 = S_{i=1}^n (w_i T x_i) \tag{6.12}$$

Each fuzzy neural operator is uniquely characterized by a number of parameters: a number of inputs, number of nodes in the hidden layer (h) and an array of connections of the AND neurons as well as the OR neuron in the output layer. The connections of the AND neurons can be systematically represented in a matrix form V, while the connections of the OR neuron are collected in a single vector form w. The overall output of the fuzzy neural operator can be described as:

$$z_j = AND(x, V_j), j = 1, 2, \ldots, h, y = OR(z, w). \tag{6.13}$$

Construction of HHCI Tree

Construction method of the HHCI tree is similar to the one used by GP or PIPE algorithm. The only difference is that a specified data structure (parameters) should be embedded into the node of the tree according to the type of the function operators selected.

Fitness Functions

A fitness function maps program to scalar, real-valued fitness values that reflect the program' performances on a given task. Firstly the fitness functions should be seen as error measures, i.e., MSE or $RMSE$. A secondary non-user-defined objective for which algorithm always optimizes programs is program size as measured by the number of nodes. Among programs with equal fitness smaller ones are always preferred.

Optimal Design

The hierarchical structure is created and optimized by using tree-structure based learning algorithms. The fine turning of the parameters encoded in the structure can be accomplished by using a set of algorithms, i.e., genetic algorithm, evolutionary strategy, evolutionary programming, differential evolution, particle swarm optimization, ant colony optimization, random search etc.

The proposed method interleaves both structure and parameter optimizations. Starting with random structures and related parameters, it first tries to improve the hierarchical structure and then as soon as an improved structure is found, it fine tunes its parameters. It then goes back to improve the structure again and, provided it finds a better structure, it again fine tunes the rules' parameters. This loop continues until a satisfactory solution (hierarchical model) is found or a time limit is reached.

The advantages of the general framework lie in that,

- For a given data set, the hierarchical structure of the HHCI models can be evolved by using tree-structure based evolutionary algorithms automatically. This is really important for the inherent hierarchical complex systems.
- It is difficult for modeling problem with many input variables, the proposed HHCI models can select the important input variables (features) automatically.
- The HHCI is a data driven automatic modeling technique, in whole process of the modeling, there is no need for prior knowledge about the plant.

6.3   Static and Dynamical Models

In Chapters 2,3 and 4 we have illustrated several variants of static hybrid computational intelligence models. These models are more efficient than a single computational intelligence model, i.e., neural network models, fuzzy system models etc. because these hybrid computational intelligence models utilize the advantages of the single models. The limitation of the static models is that the model cannot be used when the states of the system to be modeled are changed over time.

Therefore in Chapter 5, we focused on an inverse problem, the inference of a system of difference equations (DE) or a system of ordinary differential equations (ODE), from the observed time-series data using tree-structure based evolutionary algorithms. The methods can be named as reverse engineering of dynamical systems in general. It has widely applications in scientific computation, finding Free-Form Natural Laws [291], biochemical modeling and finance engineering. But it is not necessarily easy, because the appropriate form of the ODE (i.e., the order and terms) is not known in advance. We also illustrated an additive tree based evolutionary algorithm for discovery of structure forms and parameters of the system of ODE or DE. Experimental results on biochemical systems modeling and gene networks illustrate that the proposed methods are more efficient and faster than the conventional inference methods.

But there are still many problems to be addressed. The first problem is that we really need the topology, the equations, or the parameters for modeling the biochemical networks. How can we know the inferred solutions are correct or satisfy the actual facts of the biochemical process? The other problem is that the parameters determine the variables and not the other way round. If many sets of parameters can fit the same data, we must specify which one is best. To avoid reinventing the wheel, we would like to know which methods work best.

References

1. Bäck, T., Fogel, D.B., Michalewicz, Z. (eds.): Evolutionary Computation 1: Basic Algorithms and Operators. Institut of Physics Publishing (2000a)
2. Bäck, T., Fogel, D.B., Michalewicz, Z. (eds.): Evolutionary Computation 2: Advanced Algorithms and Operators. Institute of Physics Publishing (2000b)
3. Clerc, M., Kennedy, J.: The particle swarm-explosion, stability, and convergence in a multidimensional complex space. IEEE Transactions on Evolutionary Computation 6(1), 58–73 (2002)
4. Dorigo, M., Maniezzo, V., Colorni, A.: Ant system: optimization by a colony of cooperating agents. IEEE Transactions on Systems, Man, and Cybernetics-Part B 26(1), 29–41 (1996)
5. Dorigo, M., Stützle, T.: Ant Colony Optimization. MIT Press, Cambridge (2004)
6. Goldberg, D.E.: Genetic Algorithms in search, optimization, and machine learning. Addison-Wesley Publishing Corporation, Inc., Reading (1989)
7. Kennedy, J., Eberhart, R.: Swarm intelligence. Morgan Kaufmann Publishers, Inc., San Francisco (2001)
8. Kennedy, J., Mendes, R.: Population structure and particle swarm performance. In: Proceeding of IEEE conference on Evolutionary Computation, pp. 1671–1676 (2002)
9. Liu, H., Abraham, A.: Fuzzy Turbulent Particle Swarm Optimization. In: Proceeding of the 5th International Conference on Hybrid Intelligent Systems, Brazil. IEEE CS Press, USA (2005)
10. Shi, Y.H., Eberhart, R.C.: Fuzzy adaptive particle swarm optimization. In: Proceedings of IEEE International Conference on Evolutionary Computation, pp. 101–106 (2001)
11. Eberhart, R.C., Shi, Y.: Comparing inertia weights and constriction factors in particle swarm optimization. In: Proceedings of IEEE International Congress on Evolutionary Computation, pp. 84–88 (2002)
12. Orosz, J.E., Jacobson, S.H.: Analysis of static simulated annealing algorithms. Journal of Optimization theory and Applications 115(1), 165–182 (2002)
13. Parsopoulos, K.E., Vrahatis, M.N.: On the computation of all global minimizers through particle swarm optimization. IEEE Transactions on Evolutionary Computation 8(3), 211–224 (2004)
14. Triki, E., Collette, Y., Siarry, P.: A theoretical study on the behavior of simulated annealing leading to a new cooling schedule. European Journal of Operational Research 166, 77–92 (2005)
15. Ochoa, A., Muhlenbein, H., Soto, M.: A Factorized Distribution Algorithm Using Single Connected Bayesian Networks. In: Deb, K., Rudolph, G., Lutton, E., Merelo, J.J., Schoenauer, M., Schwefel, H.-P., Yao, X. (eds.) PPSN 2000. LNCS, vol. 1917, pp. 787–796. Springer, Heidelberg (2000)
16. Pelikan, M., Goldberg, D.E., Cantu-Paz, E.: BOA: The Bayesian Optimization Algorithm. In: Proceedings of the Genetic and Evolutionary Computation Conference (GECCO 1999), vol. I, pp. 525–532 (1999)
17. Rudlof, S., Koppen, M.: Stochastic Hill-Climbing with Learning by Vectors of Normal Distributions, Nagoya, Japan (1996)
18. Larranaga, P., Lozano, J.A.: Estimation of Distribution Algorithms: A New Tool for Evolutionary Computation. Kluwer Academic Publishers, Dordrecht (2001)

19. Bosman, P.A.N., Thierens, D.: Expanding from Discrete to Continuous EDAs: The IDEA. In: Deb, K., Rudolph, G., Lutton, E., Merelo, J.J., Schoenauer, M., Schwefel, H.-P., Yao, X. (eds.) PPSN 2000. LNCS, vol. 1917, pp. 767–776. Springer, Heidelberg (2000)
20. Tsutsui, S., Pelikan, M., Goldberg, D.E.: Evolutionary Algorithm Using Marginal Histogram Models in Continuous Domain. In: Proceedings of the 2001 Genetic and Evolutionary Computation Conference Workshop, San Francisco, CA, pp. 230–233 (2001)
21. Quinlan, J.R.: Induction of Decision Trees. Machine Learning 1, 81–106 (1986)
22. Quinlan, J.R.: C4.5: Programs for Machine Learning. Morgan Kaufmann, San Francisco (1993)
23. Brieman, L., Friedman, J., Olshen, R., Stone, C.: Classification of Regression Trees. Wadsworth Inc., Belmont (1984)
24. Baluja, S.: Population-based incremental learning: A method for integrating genetic search based function optimization and competitive learning. Technical report, Carnegie Mellon University, CMU-CS-94-163 (1994)
25. Ventresca, M., Tizhoosh, H.R.: A diversity maintaining population-based incremental learning algorithm. Information Sciences 178, 4038–4056 (2008)
26. Rastegar, R., Hariri, A.: The population-based incremental learning algorithm converges to local optima. Neurocomputing 69(13-15), 1772–1775 (2006)
27. Rastegar, R., Hariri, A., Mazoochi, M.: A convergence proof for the population-based incremental learning algorithm. In: Seventeenth IEEE International Conference on Tools with Artificial Intelligence, pp. 387–391 (2005)
28. Hopfield, J.J.: Neural networks and physical systems with emergent collective computational abilities. Proceedings of the National Academy of Sciences of the USA 79(8), 2554–2558 (1982)
29. Sebag, M., Ducoulombier, A.: Extending population-based incremental learning to continuous search spaces. In: Eiben, A.E., Bäck, T., Schoenauer, M., Schwefel, H.-P. (eds.) PPSN 1998. LNCS, vol. 1498, pp. 418–427. Springer, Heidelberg (1998)
30. Koza, J.R.: Genetic Programming. The MIT Press, Cambridge (1992)
31. Bonabeau, E., Dorigo, M., Theraulaz, G.: Swarm Intelligence: From Natural to Artificial Systems. Oxford University Press, New York (1999)
32. Dorigo, M., Stützle, T.: Ant Colony Optimization. MIT Press, Cambridge (2004)
33. Gambardella, L.M., Dorigo, M.: Ant-Q: A reinforcement learning approach to the traveling salesman problem. In: Proceedings of the 11th International Conference on Machine Learning, pp. 252–260 (1995)
34. Stützle, T., Hoo, H.H.: MAX-MIN ant system. Future Generation Computer Systems 16, 889–914 (2000)
35. McCulloch, W.S., Pitts, W.: A Logical Calculus of the ideas immanent in the nervous activity. Bull. Math. Biophys. 5, 115 (1943)
36. Rosenblatt, F.: The perceptron: A probabilistic model for information storage and organization in the brain. Phych. Rev. 65 (1958)
37. Minsky, M., Papert, S.: Perceptron: An introduction to computational geometry. MIT Press, Cambridge (1969)
38. Steinbush, K.: The Learning Matrix. Kybernetik (Biol. Cybern.), 36–45 (1961)
39. Amari, S.I.: Characteristics of randomly connected threshold-element network system. Proc. of IEEE 59, 35–47 (1971)
40. Fukushima, K.: Cognition: A self-organizing multilayer neural network. Biol. Cybern. 20, 121–136 (1975)

41. Yao, X.: Evolving artificial neural networks. Proceedings of IEEE 87, 1423–1447 (1999)
42. David, L.: Handbook of genetic algorithms. Van Nostrand Reinhold, New York (1991)
43. Michalewicz, Z.: Genetic algorithm+Data Structure=Evolutionary Programming. Springer, Berlin (1992)
44. Fogel, D.B., Ghozeil, A.: A note on representation and variation operators. IEEE Trans. on Evolutionary Computation 1, 159–161 (1997)
45. Montana, D.J., et al.: Training feedforward neural networks using genetic algorithms. In: Proc. of the Third Int. Conf. on Genetic Algorithms, San Mateo, pp. 379–384 (1989)
46. Sexton, R.S., et al.: Toward global optimization of neural networks: a comparison of genetic algorithm and backpropagation. Decision Support Systems 22, 171–186 (1998)
47. Porto, V.W., et al.: Alternative neural network training methods. IEEE Expert 10, 16–22 (1995)
48. Saravanan, N., et al.: A comparison of methods for self-adaptation in evolutionary algorithms. BioSystems 36, 157–166 (1995)
49. Sexton, R.S., et al.: Comparative evaluation of genetic algorithm and backpropagation for training neural networks. Information Science 129, 45–59 (2000)
50. Tan, Y., et al.: Neural network based direct optimizing predictive control with on-line PID gradient optimization. Journal of Intelligent Automation and Soft Computing (2000)
51. Takagi, T., Sugeno, M.: Fuzzy identification of systems and its applications to modeling and control. IEEE Trans. Syst., Man, Cybern. 15, 116–132 (1985)
52. Zeng, X.J., Singh, M.G.: Approximation theory of fuzzy system - MIMO case. IEEE Trans. on Fuzzy Systems 3(2), 219–235 (1995)
53. Zeng, X.J., Singh, M.G.: Approximation theory of fuzzy systems - SISO case. IEEE Trans. on Fuzzy Systems 2(2), 162–176 (1994)
54. Wang, L.X.: Universal approximation by hierarchical fuzzy systems. Fuzzy Sets and Systems 93, 223–230 (1998)
55. Huwendiek, O., Brockmann, W.: Function approximation with decomposed fuzzy systems. Fuzzy Sets and Systems 101, 273–286 (1999)
56. Hiroaki, K., et al.: Functional completeness of hierarchical fuzzy modeling. Information Science 110(1-2), 51–60 (1998)
57. Ying, H.: General SISO Takagi–Sugeno Fuzzy Systems with Linear Rule Consequent are Universal Approximators. IEEE Transactions on Fuzzy Systems 6(4), 582–587 (1998)
58. Omatu, S., et al.: Neuro-Control and its Applications. Springer, Heidelberg (1996)
59. Fahlman, S.E., et al.: The cascade-correlation learning architecture. Advances in Neural Information Processing Systems 2, 524–532 (1990)
60. Nadal, J.P.: Study of a growth algorithm for a feedforward network. International Journal of Neural Systems 1, 55–59 (1989)
61. Setiono, R., et al.: Use of a quasi-newton method in a feedforward neural network construction algorithm. IEEE Trans. on Neural Networks 6, 273–277 (1995)
62. Angeline, P.J., et al.: An evolutionary algorithm that constructs recurrent neural networks. IEEE Trans. on Neural Networks 5, 54–65 (1994)
63. Yao, X., et al.: A new evolutionary system for evolving artificial neural networks. IEEE Trans. on Neural Networks 8, 694–713 (1997)
64. Yao, X.: Evolving artificial neural networks. Proceedings of the IEEE 87, 1423–1447 (1999)

65. Stanley, K.O., Miikkulainen, R.: Evolving neural networks through augmenting topologies. Evolutionary Computation 10, 99–127 (2002)
66. Zhang, B.T., et al.: Evolutionary induction of sparse neural trees. Evolutionary Computation 5, 213–236 (1997)
67. Peter Zhang, G.: Time series forcasting using a hybrid ARIMA and neural network model. Neurocomputing 50, 159–175 (2003)
68. Working Group on Data Modeling Benchmark, Standard Committee of IEEE Neural Network Council, http://neural.cs.nthu.edu.tw/jang/benchmark/
69. Sugeno, M., et al.: A fuzzy-logic approach to qualitative modeling. IEEE Trans. Fuzzy Syst. 1, 7–31 (1993)
70. Russo, M., et al.: Generic fuzzy learning. IEEE Trans. Evol. Comput. 4, 259–273 (2000)
71. Lin, Y., et al.: Using fuzzy partitions to create fuzzy systems from input-output data and set initial weights in a fuzzy neural network. IEEE Trans. Fuzzy Syst. 5, 614–621 (1997)
72. Delgado, M., et al.: Fuzzy clustering-based rapid prototyping for fuzzy rule-based modeling. IEEE Trans. Fuzzy Syst. 5, 223–233 (1997)
73. Kukolj, D., et al.: Design of adaptive Takagi-Sugeno-Kang fuzzy models. Applied Soft Computing 2, 89–103 (2002)
74. Yao, X.: Evolving artificial neural networks. Proceedings of the IEEE 87, 1423–1447 (1999)
75. Oliver, N.: Nonlinear System Identification - From Classical Approaches to Neural Networks and Fuzzy Models. Springer, Heidelberg (2001)
76. Chebrolu, S., Abraham, A., Thomas, J.P.: Feature Detection and Ensemble Design of Intrusion Detection Systems. Computers and security (in press)
77. Salustowicz, R.P., Schmidhuber, J.: Probabilistic Incremental Program Evolution. Evolutionary Computation 2(5), 123–141 (1997)
78. Barbara, D., Couto, J., Jajodia, S., Wu, N.: ADAM: A Testbed for Exploring the Use of Data Mining in Intrusion Detection. SIGMOD Record 30(4), 15–24 (2001)
79. Lee, W., Stolfo, S., Mok, K.: A Data Mining Framework for Building Intrusion Detection Models. In: Proceedings of the IEEE Symposium on Security and Privacy (1999)
80. Debar, H., Becker, M., Siboni, D.: A Neural Network Component for an Intrusion Detection System. In: Proceedings of the IEEE Computer Society Symposium on Research in Security and Privacy (1992)
81. Friedman, J.H.: Multivariate Adaptative Regression Splines. Annals of Statistics 19 (1991)
82. Mukkamala, S., Sung, A.H., Abraham, A.: Intrusion Detection Using Ensemble of Soft Computing Paradigms. In: Third International Conference on Intelligent Systems Design and Applications. Advances in Soft Computing, pp. 239–248. Springer, Heidelberg (2003)
83. Mukkamala, S., Sung, A.H., Abraham, A.: Modeling Intrusion Detection Systems Using Linear Genetic Programming Approach. In: Orchard, B., Yang, C., Ali, M. (eds.) IEA/AIE 2004. LNCS (LNAI), vol. 3029, pp. 633–642. Springer, Heidelberg (2004)
84. Mukkamala, S., Sung, A.H., Abraham, A., Ramos, V.: Intrusion Detection Systems Using Adaptive Regression Splines. In: Seruca, I., Filipe, J., Hammoudi, S., Cordeiro, J. (eds.) 6th International Conference on Enterprise Information Systems, ICEIS 2004, Portugal, vol. 3, pp. 26–33 (2004), ISBN 972-8865-00-7

85. Shah, K., Dave, N., Chavan, S., Mukherjee, S., Abraham, A., Sanyal, S.: Adaptive Neuro-Fuzzy Intrusion Detection System. In: IEEE International Conference on Information Technology: Coding and Computing (ITCC 2004), USA, vol. 1, pp. 70–74. IEEE Computer Society, Los Alamitos (2004)
86. Mukkamala, S., Sung, A.H.: Feature Selection for Intrusion Detection Using Neural Networks and Support Vector Machines. Journal of the Transportation Research Board of the National Academics, Transportation Research Record No 1822, 33–39 (2003)
87. KDD cup 99, http://kdd.ics.uci.edu/database/kddcup99/kddcup.data_10_percent.gz
88. Working Group on Data Modeling Benchmark, Standard Committee of IEEE Neural Network Council, http://neural.cs.nthu.edu.tw/jang/benchmark/
89. Box, G.E.P.: Time series analysis, forecasting and control. Holden Day, San Francisco (1970)
90. Tong, R.M.: The evaluation of fuzzy models derived from experimental data. Fuzzy Sets and Systems 4, 1–12 (1980)
91. Pedtycz, W.: An identification algorithm in fuzzy relational systems. Fuzzy Sets and Systems 13, 153–167 (1984)
92. Xu, C.W., et al.: Fuzzy model identification and self-learning for dynamic systems. IEEE Trans. on Systems, Man and Cybernetics 17, 683–689 (1987)
93. Sugeno, M., et al.: Linguistic modelling based on numerical data. In: Proceedings of the IFSA 1991 (1991)
94. Surmann, H., et al.: Self-organising and genetic algorithm for an automatic design of fuzzy control and decision systems. In: Proceedings of the FUFIT's 1993, pp. 1079–1104 (1993)
95. Sugeno, M., et al.: A fuzzy-logic approach to qualitative modeling. IEEE Trans. Fuzzy Syst. 1, 7–31 (1993)
96. Lee, C.C., et al.: A combined approach to fuzzy model identification. IEEE Trans. on Systems, Man and Cybernetics 24, 736–744 (1994)
97. Hauptmann, W.: A neural net topology for bidirectional fuzzy-neuro transformation. In: Proceedings of the IEEE International Conference on Fuzzy Systems, pp. 1511–1518 (1995)
98. Lin, Y., et al.: A new approach to fuzzy-neural system modelling. IEEE Trans. on Fuzzy Systems 3, 190–198 (1995)
99. Nie, J.: Constructing fuzzy model by self-organising counterpropagation network. IEEE Trans. on Systems, Man and Cybernetics 25, 963–970 (1995)
100. Jang, J.S.R., et al.: Neuro-fuzzy and soft computing: a computational approach to learning and machine intelligence. Prentice-Hall, Upper saddle River (1997)
101. Kasabov, N., et al.: FuNN/2 - A fuzzy neural network architecture for adaptive learning and knowledge acquisition. Information Science 101, 155–175 (1997)
102. Kim, J., et al.: HyFIS: adaptive neuro-fuzzy inference systems and their application to nonlinear dynamical systems. Neural Networks 12, 1301–1319 (1999)
103. Cho, K.B., et al.: Radial basis function based adaptive fuzzy systems their application to system identification and prediction. Fuzzy Sets and Systems 83, 325–339 (1995)
104. Wang, L.X., et al.: Generating fuzzy rules by learning from examples. IEEE Trans. on Systems, Man and Cybernetics 22, 1414–1427 (1992)
105. Rojas, I., et al.: Time series analysis using normalized PG-RBF network with regression weights. Neurocomputing 42, 167–285 (2002)
106. Kim, D., et al.: Forecasting time series with genetic fuzzy predictor ensembles. IEEE Trans. Fuzzy Systems 5, 523–535 (1997)

107. Theodossiou, P.: The stochastic properties of major Canadian exchange rates. The Financial Review 29(2), 193–221 (1994)
108. So, M.K.P., Lam, K., Li, W.K.: Forecasting exchange rate volatility using autoregressive random variance model. Applied Financial Economics 9, 583–591 (1999)
109. Hsieh, D.A.: Modeling heteroscedasticity in daily foreign-exchange rates. Journal of Business and Economic Statistics 7, 307–317 (1989)
110. Chappel, D., Padmore, J., Mistry, P., Ellis, C.: A threshold model for French franc/Deutsch mark exchange rate. Journal of Forecasting 15, 155–164 (1996)
111. Refenes, A.N.: Constructive learning and its application to currency exchange rate forecasting. In: Trippi, R.R., Turban, E. (eds.) Neural Networks in Finance and Investing: Using Artificial Intelligence to Improve Real-World Performance, pp. 777–805. Probus Publishing Company, Chicago (1993)
112. Refenes, A.N., Azema-Barac, M., Chen, L., Karoussos, S.A.: Currency exchange rate prediction and neural network design strategies. Neural Computing and Application 1, 46–58 (1993)
113. Yu, L., Wang, S., Lai, K.K.: Adaptive Smoothing Neural Networks in Foreign Exchange Rate Forecasting. In: Sunderam, V.S., van Albada, G.D., Sloot, P.M.A., Dongarra, J. (eds.) ICCS 2005. LNCS, vol. 3516, pp. 523–530. Springer, Heidelberg (2005)
114. Yu, L., Wang, S., Lai, K.K.: A novel nonlinear ensemble forecasting model incorporating GLAR andANN for foreign exchange rates. Computers & Operations Research 32, 2523–2541 (2005)
115. Wang, W., Lai, K.K., Nakamori, Y., Wang, S.: Forecasting Foreign Exchange Rates with Artificial Neural Networks: A Review. International Journal of Information Technology & Decision Making 3(1), 145–165 (2004)
116. Chen, Y., Yang, B., Dong, J.: Nonlinear System Modeling via Optimal Design of Neural Trees. International Journal of Neural Systems 14(2), 125–137 (2004)
117. Yao, J.T., Tan, C.L.: A case study on using neural networks to perform technical forecasting of forex. Neurocomputing 34, 79–98 (2000)
118. Sastry, K., Goldberg, D.E.: Probabilistic model building and competent genetic programming. In: Riolo, R.L., Worzel, B. (eds.) Genetic Programming Theory and Practise, ch. 13, pp. 205–220. Kluwer, Dordrecht (2003)
119. http://fx.sauder.ubc.ca/
120. Yu, L., Wang, S., Lai, K.K.: Adaptive Smoothing Neural Networks in Foreign Exchange Rate Forecasting. In: Sunderam, V.S., van Albada, G.D., Sloot, P.M.A., Dongarra, J. (eds.) ICCS 2005. LNCS, vol. 3516, pp. 523–530. Springer, Heidelberg (2005)
121. Abraham, A.: Analysis of Hybrid Soft and Hard Computing Techniques for Forex Monitoring Systems. In: IEEE International Conference on Fuzzy Systems (IEEE FUZZ 2002), 2002 IEEE World Congress on Computational Intelligence, pp. 1616–1622. IEEE Press, Los Alamitos (2002)
122. Abraham, A., Chowdury, M., Petrovic-Lazerevic, S.: Australian Forex Market Analysis Using Connectionist Models. Management Journal of Management Theory and Practice 29, VIII, 18–22 (2003)
123. Zhao, W., Chellappa, R., Rosenfeld, A., Phillips, P.J.: Face Recognition: A literature survey. Technical Report CART-TR-948. University of Maryland (August 2002)
124. Chellappa, R., Wilson, C.L., Sirohey, S.: Human and machine recognition of faces: A survey. Proc. IEEE 83(5), 705–740 (1995)
125. Er, M.J., Wu, S., Lu, J., Toh, H.L.: Face recognition with radial basis function (RBF) neural networks. IEEE Trans. Neural Netw. 13(3), 697–710 (2002)

126. Yang, F., Paindavoine, M.: Implementation of an RBF neural network on embedded systems: Real-time face tracking and identity verification. IEEE Trans. Neural Netw. 14(5), 1162–1175 (2003)
127. Pan, Z., Adams, R., Bolouri, H.: Image redundancy reduction for neural network classification using discrete cosine transforms. In: Proc. IEEE-INNS-ENNS Int. Joint Conf. Neural Networks, Como, Italy, vol. 3, pp. 149–154 (2000)
128. Sorwar, G., Abraham, A., Dooley, L.: Texture Classification Based on DCT and Soft Computing. In: The 10th IEEE International Conference on Fuzzy Systems, FUZZ-IEEE 2001, vol. 2, pp. 545–548. IEEE Press, Los Alamitos (2001)
129. Valentin, D., Abdi, H., Toole, A.J.O., Cottrell, G.W.: Connectionist models of face processing: A survey. Pattern Recognit. 27, 1209–1230 (1994)
130. Su, H., Feng, D., Zhao, R.-C.: Face Recognition Using Multi-feature and Radial Basis Function Network. In: Proc. of the Pan-Sydney Area Workshop on Visual Information Processing (VIP 2002), Sydney, Australia, pp. 183–189 (2002)
131. Topon, K.P., Hitoshi, I.: Gene Selection for Classification of Cancers using Probabilistic Model Building Genetic Algorithm. BioSystems 82(3), 208–225 (2005)
132. Hong, J.-H., Cho, S.-B.: The classification of cancer based on DNA microarray data that uses diverse ensemble genetic programming. Artificial Intelligence in Medicine 36, 43–58 (2006)
133. Asyali, M.H., Colak, D., Demirkaya, O., Inan, M.S.: Gene Expression Profile Classification: A Review. Current Bioinformatics 1, 55–73 (2006)
134. Díaz-Uriarte, R., de Andrés, S.A.: Gene selection and classification of microarray data using random forest. BMC Bioinformatics 7, 3 (2006)
135. Golub, T.R., Slonim, D.K., Tamayo, P., Huard, C., Gaasenbeek, J.P., Mesirov, J., Coller, H., Loh, M.L., Downing, J.R., Caligiuri, M.A., Bloomfield, C.D., Lander, E.: Molecular Classification of Cancer: Class Discovery and Class Prediction by Gene Expression Monitoring. Science 286, 531–537 (1999)
136. Kennedy, J., Eberhart, R.: Particle Swarm Optimization. In: Proc. IEEE Int. Conf. on Neural Networks, Perth, pp. 1492–1948 (1995)
137. Frohlich, H., Chapelle, O., Scholkopf, B.: Feature Selection for Support Vector Machines by Means of Genetic Algorithms. In: 15th IEEE International Conference on Tools with Artificial Intelligence, p. 142 (2003)
138. Chen, X.-w.: Gene Selection for Cancer Classification Using Bootstrapped Genetic Algorithms and Support Vector Machines. In: IEEE Computer Society Bioinformatics Conference, p. 504 (2003)
139. Nguyen, H.-N., Ohn, S.-Y., Park, J., Park, K.-S.: Combined Kernel Function Approach in SVM for Diagnosis of Cancer. In: Wang, L., Chen, K., S. Ong, Y. (eds.) ICNC 2005. LNCS, vol. 3610, pp. 1017–1026. Springer, Heidelberg (2005)
140. Alon, U., Barkai, N., Notterman, D., Gish, K., Ybarra, S., Mack, D., Levine, A.: Broad Patterns of Gene Expression Revealed by Clustering Analysis of Tumor and Normal Colon Tissues Probed by Oligonucleotide Arrays. Proceedings of National Academy of Sciences of the United States of American 96, 6745–6750 (1999)
141. Klein, P.: Prediction of protein structural class by discriminant analysis. Biochim. Biophys. Acta 874, 205–215 (1986)
142. Chou, K.-C., Zhang, C.-T.: A correlation-coefficient method to predicting protein-structural classes from amino acid compositions. European Journal of Biochemistry 207(2), 429–433 (1992)
143. Gromiha, M.M., Ponnuswamy, P.K.: Prediction of protein secondary structures form their hydrophobic characteristics. Int. J. Pept. Protein Res. 45, 225–240 (1995)

144. Bu, W.S., Feng, Z.P., Zhang, Z., Zhang, C.T.: Prediction of protein structural classes based on amino acid index. Eur. J. Biochem. 266, 1043–1049 (1999)

145. Wang, Z.X., Yuan, Z.: How good is prediction of protein structural class by the component-coupled method. Proteins 38, 165–175 (2000)

146. Kumarevel, T.S., Gromiha, M.M., Ponnuswamy, M.N.: Structural class prediction: an application of residue distribution along the sequence. Biophys. Chem. 88, 81–101 (2000)

147. Cai, Y.D., Chou, K.C.: Predicting subcellular localization of proteins by hybridizing functional domain composition and pseudo-amino acid composition. J. Cell. Biochem. 91(6), 1197–1203 (2004)

148. Chris, H.Q.D., Dubchak, I.: Multi-class protein fold recognition using support vector machine and neural networks. Bioinformatics 17(4), 349–358 (2001)

149. Du, Q.-S., Jiang, Z.-Q., He, W.-Z.: Amino Acid Principal Component Analysis (AAPCA) and its Application in Protein Structural Class Prediction. Journal of Biomolecular Structure & Dynamics 23(6), 635–640 (2006)

150. Cheng, J., Baldi, P.: A Machine Learning Information Retrieval Approach to Protein Fold Recognition. Bioinformatics 22(12), 1456–1463 (2006)

151. Gromiha, M.M., Thangakani, A.M., Selvaraj, S.: FOLD-RATE: prediction of protein folding rates from amino acid sequence. Nucleic Acid Res. 34, 70–74 (2006)

152. Dubchak, I.: Prediction of Protein Folding Class from Amino Acid Composition. Proteins: Struct, Funct, and Genet 16, 79–91 (1993)

153. Gromiha, M.M.: A simple method for predicting transmembrane alpha helices with better accuracy. Protein 12, 557–561 (2002)

154. Shen, H.B., Yang, J., Liu, X.J.: Using supervised fuzzy clustering to predict protein structural classes. Biochem. 334(2), 577–581 (2005)

155. Chen, Y., Abraham, A., Yang, B.: Feature Selection and Classification using Flexible Neural Tree. Neurocomputing 70, 305–313 (2006)

156. Chen, Y., Jiang, S., Abraham, A.: Face Recognition Using DCT and Hybrid Flexible Tree. In: Proc. of the International Conference on Neural Networks and Brain, pp. 1459–1463 (2005)

157. Chen, Y., Peng, L., Abraham, A.: Gene Expression Profile Using Flexible Neural Tresss. In: Corchado, E., Yin, H., Botti, V., Fyfe, C. (eds.) IDEAL 2006. LNCS, vol. 4224, pp. 1121–1128. Springer, Heidelberg (2006)

158. Chen, Y., Peng, L., Abraham, A.: Exchange Rate Forecasting Using Flexible Neural Trees. In: Wang, J., Yi, Z., Żurada, J.M., Lu, B.-L., Yin, H. (eds.) ISNN 2006. LNCS, vol. 3973, pp. 518–523. Springer, Heidelberg (2006)

159. Chen, Y., Meng, Q., Zhang, Y.: Optimal Design of Hierarchial B-Spline Networks for Nonlinear System Identification. The Journal, Dynamics of Continuous, Discrete and Impulsive Systems, Series B (2006)

160. Narendra, K.S., Parthasarathy, K.: Identification and Control of Dynamical Systems Using Neural Networks. IEEE Transaction on Neural Networks 1 (1990)

161. Abraham, A., Nath, B., Mahanti, P.K.: Hybrid Intelligent Systems for Stock Market Analysis. In: Alexandrov, V.N., Dongarra, J., Juliano, B.A., Renner, R.S., Tan, C.J.K. (eds.) ICCS-ComputSci 2001. LNCS, vol. 2074, p. 337. Springer, Heidelberg (2001)

162. Abraham, A., Philip, N.S., Nath, B., Saratchandran, P.: Performance analysis of connectionist paradigms for modeling chaotic behavior of stock indices. In: Second International Workshop on Intelligent Systems Design and Applications, Computational Intelligence and Applications, pp. 181–186. Dynamic Publishers Inc., USA (2002)

163. Abraham, A., Philip, N.S., Saratchandran, P.: Modeling chaotic behavior of stock indices using intelligent paradigms. Int. J. Neural Parallel Sci. Comput. 11(1-2), 143–160 (2003)

164. Chan, W.S., Liu, W.N.: Diagnosing shocks in stock markets of southeast Asia, Australia, and New Zealand. Math. Comput. Simulation 59(1-3), 223–232 (2002)

165. Chen, S.-H., Liao, C.-C.: Agent-based computational modeling of the stock price volume relation. Inf. Sci. 170(1), 75–100 (2005)

166. Chen, Y., Abraham, A.: Feature selection and intrusion detection using hybrid flexible neural tree. In: Wang, J., Liao, X.-F., Yi, Z. (eds.) ISNN 2005. LNCS, vol. 3498, pp. 439–444. Springer, Heidelberg (2005)

167. Chen, Y., Abraham, A., Yang, B.: Hybrid Neurocomputing for Breast Cancer Detection. In: The Fourth International Workshop on Soft Computing as Transdisciplinary Science and Technology (WSTST 2005), pp. 884–892. Springer, Heidelberg (2005)

168. Francis Tay, E.S., Cao, L.J.: Modified support vector machines in financial time series forecasting. Neurocomputing 48(1-4), 847–861 (2002)

169. Geman, S., Bienenstock, E., Doursat, R.: Neural networks and the bias/variance dilemma. Neural Comput. 4, 1–58 (1992)

170. Kim, S.H., Chun, S.H.: Graded forecasting using an array of bipolar predictions: application of probabilistic neural networks to a stock market index. Int. J. Forecast. 14(3), 323–337 (1998)

171. Kim, K.J., Han, I.: Genetic algorithms approach to feature discretization in artificial neural networks for the prediction of stock price index. Expert Syst. Appl. 19(2), 125–132 (2000)

172. Leigh, W., Modani, N., Purvis, S., Roberts, T.: Stock market trading rule discovery using technical charting heuristics. Expert Syst. Appl. 23(2), 155–159 (2002)

173. Leigh, W., Paz, M., Purvis, R.: An analysis of a hybrid neural network and pattern recognition technique for predicting short-term increases in the NYSE composite index. Omega 30(2), 69–76 (2002)

174. Leigh, W., Purvis, R., Ragusa, J.M.: Forecasting the NYSE composite index with technical analysis, pattern recognizer, neural network and genetic algorithm: a case study in romantic decision support. Decision Support Syst. 32(4), 361–377 (2002)

175. Maqsood, I., Khan, M.R., Araham, A.: Neural network ensemble method for weather forecasting. Neural Comput. Appl. 13(2), 112–122 (2004)

176. Moore, A., Schneider, J., Deng, K.: Efficient locally weighted polynomial regression predictions. In: Proceedings of the 14th International Conference on Machine Learning, pp. 236–244 (1997)

177. Nasdaq Stock MarketSM, http://www.nasdaq.comi

178. National Stock Exchange of India Limited, http://www.nseindia.comi

179. Oh, K.J., Kim, K.J.: Analyzing stock market tick data using piecewise nonlinear model. Expert Syst. Appl. 22(3), 249–255 (2002)

180. Quah, T.S., Srinivasan, B.: Improving returns on stock investment through neural network selection. Expert Syst. Appl. 17(4), 295–301 (1999)

181. Refenes, A.N., Bentz, Y., Bunn, D.W., Burgess, A.N., Zapranis, A.D.: Financial time series modelling with discounted least squares backpropagation. Neurocomputing 14(2), 123–138 (1997)

182. Sharkey, A.J.C. (ed.): Combining Artificial Neural Nets. Springer, London (1999)

183. Tsaih, R., Hsu, Y., Lai, C.C.: Forecasting S&P 500 stock index futures with a hybrid AI system. Decision Support Syst. 23(2), 161–174 (1998)

184. Van den Berg, J., Kaymak, U., Van den Bergh, W.-M.: Financial markets analysis by using a probabilistic fuzzy modelling approach. Int. J. Approximate Reasoning 35(3), 291–305 (2004)
185. Wang, Y.F.: Mining stock price using fuzzy rough set system. Expert Syst. Appl. 24(1), 13–23 (2002)
186. Mat Isa, N.A., Mashor, M.Y., Othman, N.H.: Diagnosis of Cervical Cancer using Hierarchical Radial Basis Function (HiRBF) Network. In: Yaacob, S., Nagarajan, R., Chekima, A. (eds.) Proceedings of the International Conference on Artificial Intelligence in Engineering and Technology, pp. 458–463 (2002)
187. Ferrari, S., Frosio, I., Piuri, V., Alberto Borghese, N.: Automatic Multiscale Meshing Through HRBF Networks. IEEE Trans. on Instrumentation and Measurment 54(4), 1463–1470 (2005)
188. Ahmad, Z., Zhang, J.: Bayesian selective combination of multiple neural networks for improving long-range predictions in nonlinear process modelling. Neural Comput. & Applic. 14, 78–87 (2005)
189. Chen, Y., Yang, B., Dong, J.: Nonlinear systems modelling via optimal design of neural trees. International Journal of Neural Systems 14, 125–138 (2004)
190. Chen, Y., Yang, B., Dong, J., Abraham, A.: Time-series forcasting using flexible neural tree model. Information Science 174(3/4), 219–235 (2005)
191. Sastry, K., Goldberg, D.E.: Probabilistic model building and competent genetic programming. In: Riolo, R.L., Worzel, B. (eds.) Genetic Programming Theory and Practise, ch. 13, pp. 205–220. Kluwer, Dordrecht (2003)
192. Storn, R., Price, K.: Differential evolution - a simple and efficient adaptive scheme for global optimization over continuous spaces. Technical report, International Computer Science Institute, Berkley (1995)
193. Price, K.: Differential Evolution vs. the Functions of the 2nd ICEO. In: Proceedings of 1997 IEEE International Conference on Evolutionary Computation (ICEC 1997), Indianapolis, USA, pp. 153–157 (1997)
194. Chen, Y., Abraham, A., Yang, B.: Hybrid Neurocomputing for Breast Cancer Detection. In: The Fourth International Workshop on Soft Computing as Transdisciplinary Science and Technology (WSTST 2005), pp. 884–892. Springer, Heidelberg (2005)
195. Chen, Y., Yang, B., Dong, J.: Evolving Flexible Neural Networks using Ant Programming and PSO algorithm. In: Yin, F.-L., Wang, J., Guo, C. (eds.) ISNN 2004. LNCS, vol. 3173, pp. 211–216. Springer, Heidelberg (2004)
196. Merz, J., Murphy, P.M.: UCI repository of machine learning databases (1996), http://www.ics.uci.edu/-learn/MLRepository.html
197. Su, H., Feng, D., Zhao, R.-C.: Face Recognition Using Multi-feature and Radial Basis Function Network. In: Proc. of the Pan-Sydney Area Workshop on Visual Information Processing (VIP 2002), Sydney, Australia, pp. 183–189 (2002)
198. Huang, R., Pavlovic, V., Metaxas, D.N.: A hybrid face recognition method using markov random fields. In: ICPR 2004, pp. 157–160 (2004)
199. Birattari, M., Di Caro, G., Dorigo, M.: Toward the formal foundation of Ant Programming. In: Dorigo, M., Di Caro, G.A., Sampels, M. (eds.) Ant Algorithms 2002. LNCS, vol. 2463, pp. 188–201. Springer, Heidelberg (2002)
200. Takagi, T., Sugeno, M.: Fuzzy identification of systems and its application to modeling and control. IEEE Trans. Syst. Man, Cybern. 15, 116–132 (1985)
201. Xu, C., Liu, Y.: Fuzzy model identification and self learning for dynamic systems. IEEE Trans. on Syst. Man, Cybern. 17, 683–689 (1987)
202. Torra, V.: Are view of the Construction of Hierarchical Fuzzy Systems. International Journal of Intelligent Systems 17, 531–543 (2002)

203. Gan, Q., Harris, C.J.: Fuzzy local linearization and logic basis function expension in nonlinear system modeling. IEEE Trans. on Systems, Man, and Cybernetics-Part B 29(4) (1999)
204. Shi, Y., Eberhart, R., Chen, Y.: Implementation of evolutionary fuzzy systems. IEEE Trans. Fuzzy Systems 7(2), 109–119 (1999)
205. Lin, C.-K., Wang, S.-D.: Fuzzy system identification using an adaptive learning rule with terminal attractors. Journal of Fuzzy Sets and Systems, 343–352 (1999)
206. Kang, S.-J., Woo, C.-H., Hwang, H.-S., Woo, K.B.: Evolutionary design of fuzzy rule base for nonlinear system modeling and control. IEEE Trans. Fuzzy Systems 8(1), 37–45 (2000)
207. Huang, Y.-P., Wang, S.-F.: Designing a fuzzy model by adaptive macroevolution genetic algorithms. Fuzzy Sets and Systems 113, 367–379 (2000)
208. Denna, M., Mauri, G., Zanaboni, A.M.: Learning fuzzy rules with tabu search an application to control. IEEE Trans. on Fuzzy Systems 7(2), 295–318 (1999)
209. Raju, G.V., Zhou, J.: Adaptive hierarchical fuzzy controller. IEEE Trans. on System, Man and Cybernetics 23(4), 973–980 (1993)
210. Wang, L.X.: Analysis and design of hierarchical fuzzy systems. IEEE Trans. Fuzzy Systems 7(5), 617–624 (1999)
211. Wang, L.X.: Universal approximation by hierarchical fuzzy systems. Fuzzy Sets and Systems 93, 223–230 (1998)
212. Huwendiek, O., Brockmann, W.: Function approximation with decomposed fuzzy systems. Fuzzy Sets and Systems 101, 273–286 (1999)
213. Hiroaki, K., et al.: Functional completeness of hierarchical fuzzy modeling. Information Science 110(1-2), 51–60 (1998)
214. Rainer, H.: Rule generation for hierarchical fuzzy systems. In: Proc. of the annual conf. of the North America Fuzzy Information Processing, pp. 444–449 (1997)
215. Sun-Yuan, K., et al.: Synergistic modeling and applications of hierarchical fuzzy neural networks. Proceedings of IEEE 87(9), 1550–1574 (1999)
216. Brown, M., Bossley, K.M., Mills, D.J., Harris, C.J.: High dimensional neurofuzzy systems: overcoming the curse of dimensionality. In: Proc. 4th Int. Conf. on Fuzzy Systems, pp. 2139–2146 (1995)
217. Wei, C., Wang, L.-X.: A note on universal approximation by hierarchical fuzzy systems. Information Science 123, 241–248 (2000)
218. Lin, L.C., Lee, G.-Y.: Hierarchical fuzzy control for C-axis of CNC tuning centers using genetic algorithms. Journal of Intelligent and Robotic Systems 25(3), 255–275 (1999)
219. Salustowicz, R.P., Schmidhuber, J.: Probabilistic Incremental Program Evolution. Evolutionary Computation 2(5), 123–141 (1997)
220. Salustowicz, R.P., Schmidhuber, J.: Evolving structured programs with hierarchical instructions and skip nodes. In: Shavlik, J. (ed.) Machine Learning: Proceedings of the Fifteenth International Conference (ICML 1998), pp. 488–496. Morgan Kaufmann Publishers, San Francisco (1998)
221. Schmidhuber, J.: On learning how to learn learning strategies. Technical Report FKI-198-94, Fakultaet fuer Informatik, Technische Universitaet, Muenchen (revised January 1995)
222. Wiering, M.A., Schmidhuber, J.: Solving POMDPs with Levin search and EIRA. In: Saitta, L. (ed.) Machine Learning: Proceedings of the Thirteenth International Conference, pp. 534–542. Morgan Kaufmann Publishers, San Francisco (1996b)
223. Baluja, S.: Population-based incremental learning: A method for integrating genetic search based function optimization and competitive learning. Technical Report CMU-CS-94-163, Carnegie Mellon University, Pittsburgh (1994)

224. Pham, D.T., Karaboga, D.: Training Elman and Jordan Networks for System Identification using genetic algorithms. Artificial Intelligence in Engineering 13, 107–117 (1999)
225. Ying, H.: Theory and application of a novel fuzzy PID controller using a simplified Takagi-Sugeno rule scheme. Information Science 123(3-4), 281–293 (2000)
226. Cramer, N.L.: A representation for the adaptive generation of simple sequential programs. In: Grefenstette, J. (ed.) Proceedings of an International Conference on Genetic Algorithms and Their Applications, pp. 183–187. Lawrence Erlbaum Associates, Hillsdale (1985)
227. Duan, J.-C., Chung, F.-L.: Multilevel fuzzy relational systems: structure and identification. Soft Computing 6, 71–86 (2002)
228. Babuska, R.: Fuzzy modeling and identification, Ph.D. Thesis, University of Delft, the Netherlands (1996)
229. Angelov, P., Filev, D.: An Approach to Online Identification of Takagi-Sugeno Fuzzy Models. IEEE Transactions on System, Man, and Cybernetics, part B - Cybernetics 34(1), 484–498 (2004)
230. Kasabov, N., Song, Q.: DENFIS: Dynamic, evolving neural-fuzzy inference system and its application for time-series prediction. IEEE Trans. on Fuzzy Systems 10, 144–154 (2002)
231. Shimojima, K., Fukuda, T., Hasegawa, Y.: Self-turning fuzzy modeling with adaptive membership function, rules, and hierarchical structure based on genetic algorithm. Fuzzy Sets and Systems 71, 295–309 (1995)
232. Babuska, R.: Fuzzy modeling and identification, Ph.D. Thesis, University of Delft, the Netherlands (1996)
233. Tachibana, K., Furuhashi, T.: A Structure Identification Method of Submodels for Hierarchical Fuzzy Modeling Using the Multiple Objective Genetic Algorithm. International Journal of Intelligent Systems 17, 495–531 (2002)
234. Chen, Y., Yang, B., Dong, J.: Automatic Design of Hierarchical TS-FS Models using Ant Programming and PSO algorithm. In: Bussler, C.J., Fensel, D. (eds.) AIMSA 2004. LNCS (LNAI), vol. 3192, pp. 285–294. Springer, Heidelberg (2004)
235. Ishibuchi, H., Murata, T., Turksen, I.B.: Single-objective and two-objective genetic algorithms for selecting linguistic rules for pattern classification problems. Fuzzy Sets Syst. 89, 135–150 (1997)
236. Ishibuchi, H., Nakashima, T., Murata, T.: Three-objective genetic based machine learning for linguistic rule extraction. Inf. Sci. 136, 109–133 (2001)
237. Anderson, E.: The Irises of the Gaspe peninsula. Bull. Amer. Iris Soc. 59, 2–5 (1935)
238. Wang, J.S., Lee, G.C.S.: Self-adaptive neuro-fuzzy inference system for classification application. IEEE Trans. Fuzzy Syst. 10, 790–802 (2002)
239. Wu, T.P., Chen, S.M.: A new method for constructing membership functions and fuzzy rules from training examples. IEEE Trans. Syst., Man, Cybern. B 29, 25–40 (1999)
240. Russo, M.: Genetic fuzzy learning. IEEE Trans. Evolut. Computat. 4, 259–273 (2000)
241. Setnes, M., Roubos, H.: GA-fuzzy modeling and classification: complexity and performance. IEEE Trans. Fuzzy Syst. 8, 509–522 (2000)
242. Roubos, J.A., Setnes, M., Abonyi, J.: Learning fuzzy classification rules from labeled data. Inform. Sci. 150(1-2), 77–93 (2003)
243. Abraham, A.: Adaptation of Fuzzy Inference System Using Neural Learning. In: Nedjah, N., et al. (eds.) Fuzzy System Engineering: Theory and Practice, ch. 3. Studies in Fuzziness and Soft Computing, pp. 53–83. Springer, Heidelberg (2005)

244. Abraham, A.: EvoNF: A Framework for Optimization of Fuzzy Inference Systems Using Neural Network Learning and Evolutionary Computation. In: The 17th IEEE International Symposium on Intelligent Control, ISIC 2002, pp. 327–332. IEEE Press, Los Alamitos (2002)

245. Andrew, H.W., et al.: System Identification using Genetic Programming. In: Proc. of 2nd Int. Conference on Adaptive Computing in Engineering Design and Control (1996)

246. Howley, B.: Genetic Programming of Near minimum Time Spacecraft Attitude Manoeuvres. In: Proc. of the 1st Annual Conference on Genetic Programming, Stanford University, pp. 98–106 (1996)

247. Dracopoulos, D.C.: Evolutionary Control of a Satellite. In: Proc. of the 2nd Annual Conference on Genetic Programming, Stanford University, pp. 77–81 (1997)

248. Dominic, S.: Chemical Process Controller Design using Genetic Programming. In: Proc. of the 3rd Annual Conference on Genetic Programming, University of Wisconsin, pp. 77–81 (1998)

249. Kumar, C.: Automatic generation of non-linear optimal control laws for broom balancing using evolutionary programming. In: Porto, V.W., Waagen, D. (eds.) EP 1998. LNCS, vol. 1447. Springer, Heidelberg (1998)

250. Angeline, P.J., Fogel, D.B.: An Evolutionary Program for the Identification of Dynamical System. In: Proc. of SPIE, Bellingham, pp. 409–417 (1997)

251. Kristinsson, K., et al.: System Identification and Control Using Genetic Algorithms. IEEE Trans. on System Man Cybernetics (5), 1033–1046 (1992)

252. Banzhaf, W., Nordin, P., et al.: Genetic Programming - An Introduction. Morgan Kaufmann, San Francisco (1998)

253. Rafal, S., Jurgen, S.: Probabilistic Incremental Program Evolution. Evolutionary Computation 2(5), 123–141 (1997)

254. Chen, Y., Kawaji, S.: Identification and Control of Nonlinear System using PIPE Algorithm. In: Proc. of Workshop on Soft Computing in Industry 1999, Muroran, Japan, pp. 60–65 (1999)

255. Chen, Y., Kawaji, S.: Evolutionary Control of Discrete-Time Nonlinear System using PIPE Algorithm. In: Proc. of Int. Conference on System, Man and Cybernetics (IEEE SMC 1999), Tokyo, pp. 1078–1083 (1999)

256. Chen, Y., Kawaji, S.: Evolving Wavelet Neural Networks for System Identification. In: Proc. of the International Conference on Electrical Engineering (ICEE 2000), Kitakyushu, Japan, pp. 279–282 (2000)

257. Chen, Y., Kawaji, S.: System Identification and Control using Probabilistic Incremental Program Evolution Algorithm. Journal of Robotics and Machatronics 12(6) (2000)

258. Chen, Y., Kawaji, S.: Evolving Neurofuzzy Systems by Hybrid Soft Computing Approaches for System Identification. International Journal of Advanced Computational Intelligence 6, 220–228 (2002)

259. Chellapilla, K., et al.: Evolving computer programs without subtree crossover. IEEE Transactions on Evolutionary Computation 1, 209–216 (1997)

260. Hu, J., et al.: RasID - Random Search for Neural Network Training. Journal of Advanced Computational Intelligence 2(2), 134–141 (1998)

261. Martyas, J.: Random Optimization. Automation and Remote Control 28, 244–251 (1965)

262. Solis, F.J., et al.: Minimization by random search techniques. Mathematics of Operations Research 6, 19–30 (1981)

263. Tsoulos, I.G., Lagaris, I.E.: Solving differential equations with genetic programming. Genetic Programming and Evolvable Machines 7(1), 1389–2576 (2006)

264. Cao, H., Kang, L., Chen, Y., Yu, J.: Evolutionary Modeling of Systems of Ordinary Differential Equations with Genetic Programming. Genetic Programming and Evolvable Machines 1(40), 309–337 (2000)

265. Iba, H., Mimura, A.: Inference of a gene regulatory network by means of interactive evolutionary computing. Information Sciences 145(3-4), 225–236 (2002)

266. Oltean, M., Dumitrescu, D.: Multi expression programming. Technical report, UBB-01-2002, Babes-Bolyai University, Cluj-Napoca, Romania, http://www.mep.cs.ubbcluj.ro

267. Grosan, C., Abraham, A., Han, S.Y.: MEPIDS: Multi-Expression Programming for Intrusion Detection System. In: Mira, J., Álvarez, J.R. (eds.) IWINAC 2005. LNCS, vol. 3562, pp. 163–172. Springer, Heidelberg (2005)

268. Andrew, H.W., et al.: System Identification using Genetic Programming. In: Proc. of 2nd Int. Conference on Adaptive Computing in Engineering Design and Control (1996)

269. Oltean, M., Grosan, C.: Evolving digital circuits using multi expression programming. In: Zebulum, R., et al. (eds.) NASA/DoD Conference on Evolvable Hardware, Seattle, June 24-26, pp. 87–90. IEEE Press, Los Alamitos (2004)

270. Chen, Y., Yang, B., Abraham, A.: Flexible Neural Trees Ensemble for Stock Index Modeling. Neurocomputing 70(4-6), 697–703 (2007)

271. Takeuchi, Y.: Global Dynamical Properties of Lotka-Volterra Systems. World Scientific, Singapore (1996)

272. Arkin, A.P., Ross, J.: Statistical construction of chemical reaction mechanisms from measured time-series. J. Phys. Chem., 970–979 (1995)

273. Iba, H.: Inference of differential equation models by genetic programming. Information Sciences 178, 4453–4468 (2008)

274. Girolami, M.: Bayesian inference for differential equations. Theoretical Computer Science 408, 4–16 (2008)

275. Qian, L.: Inference of Noisy Nonlinear Differential Equation Models for Gene Regulatory Networks Using Genetic Programming and Kalman Filtering. IEEE Transactions on Signal Processing 56(7), 3327–3339 (2008)

276. Akutsu, T., Miyano, S., Kuhara, S.: Identification of Genetic Networks from a Small Number of Gene Expression Patterns under the Boolean Network Model. In: Proc. of Pacific Symposium on Biocomputing, pp. 17–28 (1999)

277. Murphy, K., Mian, S.: Modeling gene expression data using dynamic Bayesian network. Computer Science Division, University of California, Berkeley (1999)

278. Chen, T., He, H.L., Church, G.M.: Modeling Gene Expression with Differential Equations. In: Proc. of Pacific Symposium on Biocomputing, pp. 29–40 (1999)

279. Tominaga, D., Koga, N., Okamoto, M.: Efficient Numerical Optimization Algorithm Based on Genetic Algorithm for Inverse Problem. In: Proc. of Genetic and Evolutionary Computation Conference, pp. 251–258 (2000)

280. Sakamoto, E., Iba, H.: Inferring a system of differential equations for a gene regulatory network by using genetic programming. In: Proc. Congress on Evolutionary Computation, pp. 720–726 (2001)

281. Cho, D.Y., Cho, K.H., Zhang, B.T.: Identification of biochemical networks by S-tree based genetic programming. Bioinformatics 22(13), 1631–1640 (2006)

282. Bornholdt, S.: Boolean network models of cellular regulation: prospects and limitations. J. R. Soc. Interf. 5, S85–S94 (2008)

283. Hecker, M., Lambeck, S., Toepfer, S., van Someren, E., Guthke, R.: Gene regulatory network inference: Data integration in dynamic models–review. Biosystems 96(1), 86–103 (2009)

284. Bongard, J., Lipson, H.: Automated reverse engineering of nonlinear dynamical systems. Proceedings of the National Academy of Science 104(24), 9943–9948 (2007)

285. Gennemark, P., Wedelin, D.: Efficient algorithms for ordinary differential equation model identification of biological systems. IET Syst. Biol. 1(2), 120–129 (2007)

286. Savageau, M.A.: Biochemical systems analysis: a study of function and design in molecular biology. Addison-Wesley, Reading (1976)

287. Maki, Y., Tominaga, D., Okamoto, M., Watanabe, S., Eguchi, Y.: Development of a system for the inference of large scale genetic networks. In: Pac. Symp. Biocomput., pp. 446–458 (2001)

288. Kimura, S., Ide, K., Kashihara, A.: Inference of S-system models of genetic networks using a cooperative coevolutionary algorithm. Bioinformatics 21(7), 1154–1163 (2005)

289. Kikuchi, S., et al.: Dynamic modeling of genetic networks using genetic algorithm and S-system. Bioinformatics 19, 643–650

290. Abraham, A., Jain, L., Kacprzyk, J. (eds.): Recent Advances in Intelligent Paradigms and Applications. Studies in Fuzziness and Soft Computing. Physica Verlag, Heidelberg (2002)

291. Schmidt, M., Lipson, H.: Distilling Free-Form Natural Laws from Experimental Data. Science 324, 81–85 (2009)

292. Macready, W.G., Wolpert, D.H.: The No Free Lunch theorems. IEEE Trans. on Evolutionary Computing 1(1), 67–82 (1997)

293. Abraham, A.: Meta-Learning Evolutionary Artificial Neural Networks. Neurocomputing Journal 56c, 1–38 (2004)

294. Abraham, A.: Optimization of Evolutionary Neural Networks Using Hybrid Learning Algorithms. In: IEEE 2002 Joint International Conference on Neural networks, World Congress on Computational Intelligence (WCCI 2002), Hawaii, May 12-17 (2002)

295. Abraham, A., Nath, B.: Optimal Design of Neural Nets Using Hybrid Algorithms. In: Mizoguchi, R., Slaney, J.K. (eds.) PRICAI 2000. LNCS, vol. 1886, pp. 510–520. Springer, Heidelberg (2000)

296. Mandic, D., Chambers, J.: Recurrent Neural Networks for Prediction: Learning Algorithms, Architectures and Stability. John Wiley & Sons, USA (2001)

297. Carpenter, G., Grossberg, S.: Adaptive Resonance Theory (ART). In: Arbib, M. (ed.) The Handbook of Brain Theory and Neural Networks, pp. 79–82. MIT Press, Cambridge (1995)

298. Bishop, C.M.: Neural Networks for Pattern Recognition. Oxford University Press, Oxford (1995)

299. Chen, S., Cowan, C.F.N., Grant, P.M.: Orthogonal Least Squares Learning Algorithm for Radial Basis Function Networks. IEEE Transactions on Neural Networks 2(2), 302–309 (1991)

300. Kohonen, T.: Self-Organization and Associative Memory. Springer, New York (1988)

301. Grossberg, S.: Adaptive pattern classification and universal recoding: Parallel development and coding of neural feature detectors. Biological Cybernetics 23, 121–134 (1976)

302. Zadeh, L.A.: Fuzzy Sets. Information and Control 8, 338–353 (1965)

303. Abraham, A.: Nature and Scope of AI Techniques. In: Sydenham, P., Thorn, R. (eds.) Handbook for Measurement Systems Design, pp. 893–900. John Wiley and Sons Ltd., London (2005)

304. Abraham, A.: Artificial Neural Networks. In: Sydenham, P., Thorn, R. (eds.) Handbook for Measurement Systems Design, pp. 901–908. John Wiley and Sons Ltd., London (2005)
305. Abraham, A.: Rule Based Expert Systems. In: Sydenham, P., Thorn, R. (eds.) Handbook for Measurement Systems Design, pp. 909–919. John Wiley and Sons Ltd., London (2005)
306. Abraham, A.: Evolutionary Computation. In: Sydenham, P., Thorn, R. (eds.) Handbook for Measurement Systems Design, pp. 920–931. John Wiley and Sons Ltd., London (2005)
307. Abraham, A.: Intelligent Systems: Architectures and Perspectives. In: Abraham, A., Jain, L., Kacprzyk, J. (eds.) Recent Advances in Intelligent Paradigms and Applications, ch. 1. Studies in Fuzziness and Soft Computing, pp. 1–35. Springer, Heidelberg (2002)
308. Abraham, A.: Adaptation of Fuzzy Inference System Using Neural Learning. In: Nedjah, N., et al. (eds.) Fuzzy System Engineering: Theory and Practice, ch. 3. Studies in Fuzziness and Soft Computing, pp. 53–83. Springer, Heidelberg (2005)
309. Abraham, A., Khan, M.R.: Neuro-Fuzzy Paradigms for Intelligent Energy Management. In: Abraham, A., Jain, L., Jan van der Zwaag, B. (eds.) Innovations in Intelligent Systems: Design, Management and Applications, ch. 12. Studies in Fuzziness and Soft Computing, pp. 285–314. Springer, Heidelberg (2003)
310. Abraham, A.: Neuro-Fuzzy Systems: State-of-the-Art Modeling Techniques. In: Mira, J., Prieto, A.G. (eds.) IWANN 2001. LNCS, vol. 2084, pp. 269–276. Springer, Heidelberg (2001)
311. Cordon, O., Herrera, F., Hoffmann, F., Magdalena, L.: Genetic Fuzzy Systems: Evolutionary Tuning and Learning of Fuzzy Knowledge Bases. World Scientific Publishing Company, Singapore (2001)
312. Abraham, A., Guo, H., Liu, H.: Swarm Intelligence: Foundations, Perspectives and Applications. In: Nedjah, N., Mourelle, L. (eds.) Swarm Intelligent Systems. Studies in Computational Intelligence, pp. 3–25. Springer, Heidelberg (2006)
313. Pearl, J.: Probabilistic Reasoning in Intelligent Systems: Networks of Plausible Inference. Morgan Kaufmann Publishers, San Francisco (1997)